Technosignatures for Detecting Intelligent Life in Our Universe

Scrivener Publishing
100 Cummings Center, Suite 541J
Beverly, MA 01915-6106

Astrobiology Perspectives on Life of the Universe

Series Editors: Richard Gordon and Joseph Seckbach

In his 1687 book *Principia*, Isaac Newton showed how a body launched atop a tall mountain parallel to the ground would circle the Earth. Many of us are old enough to have witnessed the realization of this dream in the launch of Sputnik in 1957. Since then our ability to enter, view and understand the Universe has increased dramatically. A great race is on to discover real extraterrestrial life, and to understand our origins, whether on Earth or elsewhere. We take part of the title for this new series of books from the pioneering thoughts of Svante Arrhenius, who reviewed this quest in his 1909 book ***The Life of the Universe as Conceived by Man from the*** *Earliest Ages to the Present Time*. The volumes in ***Astrobiology Perspectives on Life of the Universe*** will each delve into an aspect of this adventure, with chapters by those who are involved in it, as well as careful observers and assessors of our progress. Guest editors are invited from time to time, and all chapters are peer-reviewed.

Publishers at Scrivener
Martin Scrivener (martin@scrivenerpublishing.com)
Phillip Carmical (pcarmical@scrivenerpublishing.com)

Technosignatures for Detecting Intelligent Life in Our Universe

A Research Companion

Edited by

Anamaria Berea
Dept. of Data and Computational Sciences, George Mason University, Virginia, USA

Scrivener Publishing

WILEY

This edition first published 2022 by John Wiley & Sons, Inc., 111 River Street, Hoboken, NJ 07030, USA and Scrivener Publishing LLC, 100 Cummings Center, Suite 541J, Beverly, MA 01915, USA

For more information about Scrivener publications please visit www.scrivenerpublishing.com.

Wiley Global Headquarters
111 River Street, Hoboken, NJ 07030, USA

For details of our global editorial offices, customer services, and more information about Wiley products visit us at www.wiley.com.

Library of Congress Cataloging-in-Publication Data

ISBN 978-1-119-64040-0

Cover image: https://commons.wikimedia.org/wiki/File:Dyson_Sphere_Render.png
Cover design by Russell Richardson

Set in size of 11pt and Minion Pro by Manila Typesetting Company, Makati, Philippines

Printed in the USA

10 9 8 7 6 5 4 3 2 1

Contents

Preface

Technosignature is a term that encompasses the idea of searching for intelligent, technological signatures of civilizations other than ours in the universe. A term that would probably be more familiar to the larger audience would be SETI (search for extraterrestrial intelligence). Technosignatures are the technological equivalent of biosignatures and a subset of biosignatures, which represent evidence of alien life, whether it is intelligent or not. They represent any signals that can be collected from the universe, such as radio wavelengths, optical signals, and many more, that can be potential candidates as signals emitted intentionally from another part of the universe that is not Earth.

This book—*Technosignatures for Detecting Intelligent Life in Our Universe: A Research Companion*—represents a collection of works by a wide diversity of researchers who have been in this field for a long time. While there have been several books written on the subject in the past, this is the newest research that we have to date in the field, and it by no means represents all aspects of technosignatures research. But this book does give us a good lens into the current state-of-the-art in the field. At the same time, it spreads across a wide range of other fields that are tangential or are used for SETI— from history and economics to communication, photometry, statistics, data science, astrobiology, exoplanetary science, and many more.

Some of the most important debates, when searching for life beyond our planet, have been around the definitions of life and intelligence. While we don't have very accurate or universally agreed upon definitions for either of these monumentally important concepts, we can recognize, in general, what is life is and what intelligence is, especially on our planet. And, in general, we assume there is no intelligence without life, as we also consider artificial intelligence to be an extension and a byproduct of intelligent biological life. In fact, if biosignatures are signatures of life, in general, technosignatures are signatures of intelligent life, whether this life is biological, artificial, or even extinct. There are many voices in the community that are arguing that actually, in case we do find technosignatures, it is more likely that they will be some forms of artificial intelligence or signs of current or extinct civilizations.

At the same time, let's not forget what Arthur C. Clarke said, that if we do find an advanced alien civilization, it is very likely that their technology would be, for us, undistinguishable from magic. Therefore, searching for technosignatures is not a trivial pursuit, and given how big the question "Are We Alone?" is, the problem we are facing is one of the most important and perennial in science.

As our science and scientific understanding and instruments advance, so do our definitions and framing or reframing of the same problem. The way we understand life and intelligence today, as well as *potential* life and intelligence, will co-evolve with our new discoveries, both here, on our planet, and in the Universe. And research companions like this one help with the current state of the field, but will also have to be updated and changed in time.

The first chapter, "Historical Perspectives - How the Search for Technosignatures Grew Out of the Cold War", takes us into the detailed history of the field, the origins of SETI and the importance of placing this field in historical context. The second chapter, "Reading the Cosmos: What Our Science Tells Us about the Science of Another World", highlights the importance of scientific process and evolution of science per se for an advanced civilization that can develop technologies and can communicate at interstellar distances, as well as how perceptions of space and time might affect communication with possible intelligent alien life. The third chapter, "The Impact of Discovering the First Technosignature", delves into the social and societal impact of such a big discovery and how we can prepare here on Earth for disseminating such information to the public at large. The fourth chapter, "Searching for Extraterrestrial Intelligence by Locating Potential ET Communication Networks in Space", outlines in more depth the possibility of communication networks from alien life, which would imply planetary and habitability networks, as well as more advanced and sophisticated means of communication. The fifth chapter, "Habitable Mini-Earths with Black Hole Cores", proposes a new model of habitability for smaller planets, such as mini-Earths, that would have a black hole at their core. The sixth chapter, "Technosignatures in Time-Series Photometry", proposes a new method for detecting technosignatures, based on the analysis of timeseries data in the light curves of photometry of exoplanets. The seventh chapter, "Post-Detection Message Analysis and Comprehension", looks at the efforts that would be needed for the analysis, comprehension and possibly reply after such an alien message would be detected, from the point of view of information science and patterns. The eighth chapter, "Statistical Issues in the Search for Technosignatures", develops a full physical-statistical description of electromagnetic technosignatures that could possibly populate the Milky Way from a technical point of view. The ninth and last chapter, "Economics and Technosignatures: New Connections", explores the many ways in which ideas and findings from economics can be borrowed by the field of technosignatures in ways that can advance both detection and implications of detection of such a message or signal.

This collection of essays therefore represents an updated reference material for the field of technosignatures, and includes novel ideas from economics, information theory, astrophysics, statistics, social sciences, planetary sciences, and more. It is a research compendium for scientists from any interdisciplinary field who would want to get involved in this exciting area of work, and it is accessible not only to scientists but to the public at large as well.

Although not extensive, this book summarizes the multiple interdisciplinary efforts that have contributed to the field of technosignatures, particularly within the past few years. It shows how current advances in science, technology, and social sciences can support this effort and can be used as both a resource for the scientists in the field and as a reference for the public interested in the topic.

Anamaria Berea
anamaria@bmsis.org or aberea@gmu.edu
April 2022

1

Historical Perspectives: How the Search for Technosignatures Grew Out of the Cold War

Rebecca Charbonneau

Historian-in-Residence, Harvard-Smithsonian Center for Astrophysics, Charlottesville, USA

Abstract
This chapter focuses on the history of the search for extraterrestrial intelligence. It will explore the following questions: Why did SETI transition from a fringe idea to a concerted scientific effort in the 1960s? What were the different cultural, ideological, and technical approaches in the search? Why did SETI develop primarily in the US and Soviet Union as opposed to other countries? What challenges did early SETI pioneers face? In investigating these questions, I will focus especially on the collaborative efforts between US astronomer Carl Sagan and Soviet astrophysicist I.S Shklovsky as my main case study. This chapter aims to contextualize the current effort to search for technosignatures within its historical roots, with hopes this context will prompt a mindful analysis of how we utilize the artefacts from history in the ongoing search.

Keywords: SETI, technosignatures, history, history of science, radio astronomy, cold war, soviet union, Carl Sagan

1.1 Introduction

The search for and communication with extraterrestrial intelligence (SETI/CETI) [1.1] has largely been considered an internationalist scientific pursuit because of the "unifying" aspects of viewing humanity as a singular

Email: Rebecca.charbonneau@cfa.harvard.edu
Bio Links: https://rebeccacharbonneau.wordpress.com/

Anamaria Berea (ed.) Technosignatures for Detecting Intelligent Life in Our Universe: A Research Companion, (1–32) © 2022 Scrivener Publishing LLC

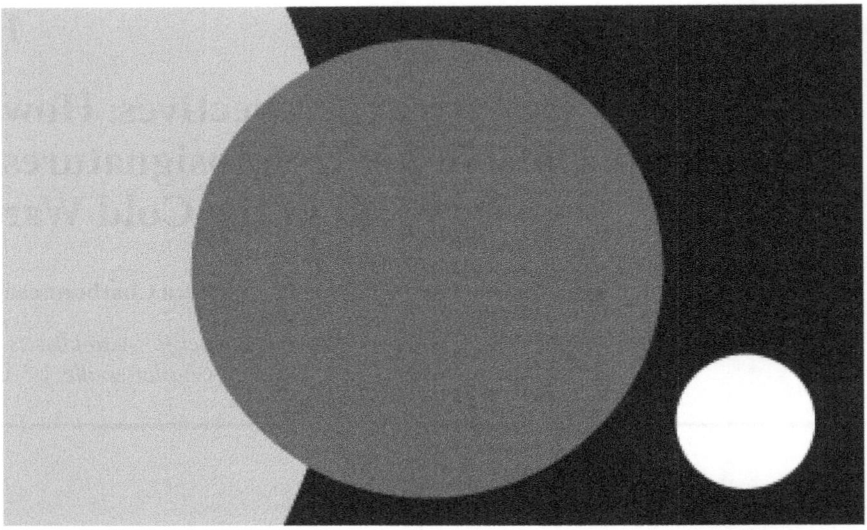

Figure 1.1 The flag of Earth, created in 1970 by James W. Cadle.

whole in a potential universe teeming with other intelligent civilizations. This hopeful scientific internationalism is reflected in much of the discourse and material culture surrounding the search for technosignatures; for example, the anti-national "flag of Earth" (Figure 1.1) flies in many places associated with the search, such as the Ohio State University Radio Observatory and the offices of the premier twenty-first century technosignature initiative, Breakthrough Listen. Indeed, many technosignature researchers, in studying the potential cultural impact of discovering extraterrestrial intelligence, have argued that the discovery of life on other worlds could possibly bring about global unity [1.2].

Despite the internationalist goals of the discipline, the early development of the search for technosignatures was deeply entrenched in the contentious geopolitics of its time, most notably in the tensions between the US and Soviet Union. Many introductory histories to SETI largely focus on significant events in SETI, such as the first SETI search in radio astronomy, Project Ozma, or introduce the major scientific contributors to the science. This chapter attempts to take a different approach; rather than focusing on key events and people, I will attempt to contextualise the origins of the science in the larger historical context. In doing so, this chapter will aim to answer the following questions: Why did CETI transition from a fringe idea to a concerted scientific effort in the 1960s? What motivated early CETI scientists to pursue their work? And why did CETI develop primarily in the US and Soviet Union as opposed to other countries? Answering these questions as they relate to one of the earliest efforts in the search for technosignatures will show that CETI was a direct product of the Cold War, primarily due to major government investment in scientific infrastructure and the influence of the Space Race on Soviet and American scientific and popular culture.

Additionally, CETI's development during the Cold War led CETI researchers to face specific challenges regarding communication and collaboration across the Iron Curtain. I will show here that although CETI scientists in the 1960s were preoccupied with finding and communicating with extraterrestrial life, they simultaneously faced great challenges meeting and communicating with one another. These barriers to communication included restrictions and censorship in publication, mail interference, travel restrictions, and ideological differences. We might describe these as difficulties transmitting and interpreting textual signatures, and I will suggest that their search for extraterrestrial intelligence was complexly related to the diverse dimensions of interterrestrial intelligence.

It is my hope this chapter will give the reader a greater appreciation for the tenacity of early CETI pioneers, especially those who contributed from the Soviet Union, as well as recognize the influence of the Cold War on the philosophical questions posed by CETI. To illustrate the crisis of communication with the "alien"—both those on other planets and on Earth—during the Cold War, I will focus especially on the interactions between US astronomer Carl Sagan and Soviet astrophysicist I.S. Shklovsky and their attempts at publishing the first popular CETI book in the mid-1960s. In exploring the history of the Cold War and relations between the Soviet Union and United States, a greater context for the development of CETI emerges, providing insight into both the cosmic and Earth-based challenges faced by CETI pioneers. Understanding that context will hopefully prompt a mindful analysis of how we utilize the artefacts from this history in the ongoing search for technosignatures.

1.2 The Extraterrestrial Life Debate Gets Technical

To understand the role of the Cold War in the development of CETI, it is first important to recognize how the intervention of technology revolutionized the approach to the question of the plurality of worlds. Up until the nineteenth century and the start of the Cold War, the question of humanity's place in the cosmos remained a purely speculative one, largely resigned to the musings of philosophers and theologians. In ancient Greece, atomist thinkers theorized on the plurality of *kosmoi* [1.3], and in early modern

Europe, physicists inspired by Copernicanism published tomes speculating on the existence of other planets, peopled just as Earth was [1.4]. These ponderings on the nature of the universe over the course of millennia are often referred to as the "extraterrestrial life debate".

In the nineteenth century, however, scientists began to explore the question of extraterrestrial life using scientific instruments such as the optical telescope, though their attempts remained largely rooted in conjecture and imagination, rather than systematic and empirical inquiry. For example, in 1906, Percival Lowell published a book titled *Mars and Its Canals*. In this book, Lowell, an American businessman and mathematician who had previously founded an astronomical observatory, asserted Mars was populated by intelligent life that had built intricate canals on the surface of the Red Planet [1.5]. Such assertions were not without some observational basis; Lowell had made countless observations of Mars in his observatory, and in 1907 published a mathematical essay in which he attempted to prove that the mean surface temperature of Mars was similar to an unseasonably warm winter day in England [1.6]. Still, even speculation rooted in some form of empiricism was inevitably unsubstantiated, and as a result, drew criticism and mockery from both the public and his scientific contemporaries. Alfred Russel Wallace, for example, published an entire book in refutation of Lowell's assertions, in which he sharply concluded: "Mars... is not only uninhabited by intelligent beings such as Mr. Lowell postulates, but is absolutely *uninhabitable*" [1.7]. Because of a hypothesis which lacked falsifiability, in addition to the eccentric characters who investigated it, such as Lowell, the extraterrestrial life debate had a dubious reputation within the scientific community in the nineteenth and early twentieth centuries. But in the mid-twentieth century there was a dramatic shift in the scientific perception of the extraterrestrial life debate, enabled by new technologies of radio communication and radar techniques.

Interest in using radio technology to communicate with extraterrestrials has been around for nearly as long as radio technology itself. In 1896, Serbian-American electrical engineer and physicist Nikola Tesla asserted that his new electrical transmission system could be used to communicate with Mars and soon after, Italian radio engineer Guglielmo Marconi claimed to have received radio signals from Mars. Although Tesla claimed to have used his radio experiments to become "the first to hear the greeting of one planet to another" in 1901, it was not until about two decades later that using radio technology to communicate with other planets became widely discussed in scientific circles [1.8]. In January 1919, the *New York Times* published an article titled "Radio to Stars, Marconi's Hope" [1.9]. It contained a summation of an interview the English journalist Harold Begbie had conducted with Marconi, during which Marconi discussed "the possibility of communicating by wireless with the stars" [1.10]. In the interview, Marconi speculated on the potential use of radio technology for interstellar communication. He said:

> Messages that I sent off ten years ago have not yet reached the nearest stars. When they arrive there why should they stop? ... That is what makes me hope for a very big thing in the future... communication with intelligences on other stars [1.11].

The very next year, Marconi announced that he was investigating signals he postulated migh have come from Mars [1.12]. The *New York Times* once again published a lengthy article on Marconi's so-called "Mars signals". In Europe, Marconi's claims faced mostly ridicule; one French newspaper, for example, publicized Marconi's discovery under the headline "Hello, Central, give me the moon", referring sardonically to Marconi's wireless telegraphy system [1.13]. Yet the signals Marconi received, which he described as "distinct but unintelligible", generated much excitement in the United States. A newspaper in Minnesota, *The Tomahawk*, published an article titled "Hello Earth! Hello!", which jumped at the possibility of

Figure 1.2 Illustrations of Tesla and Marconi in an article titled "Hello Earth! Hello!" in The Tomahawk, March 18, 1920, Image 6. Digital scan held in the Library of Congress.

communication with extraterrestrial intelligence (Figure 1.2). In the article, Marconi is quoted as having said:

> If there are any human beings on Mars I would not be surprised if they should find a means of communication with this planet. Linking of the science of astronomy with that of electricity may bring about almost anything [1.14].

Still, despite interest and international conversation, these early investigations into the use of radio technology to communicate with extraterrestrials were rooted in conjecture and imagination, not rigorous scientific investigation, not unlike those of Lowell. For example, according to *The Tomahawk*, Marconi proposed Martians may use Morse code to communicate with Earth, which demonstrates that although ideas for interplanetary radio contact were forming, there was no serious scientific investigation into the problem of communication with extraterrestrial intelligence.

The advent of radio astronomy shifted the investigation of communication with extraterrestrial intelligence from a speculative one to a truly technical one because it gave scientists the opportunity to test their theories by making strategic radio observations of the cosmos. Although the science began as early as 1931 with Karl Jansky's discovery of the cosmic sources of radio waves [1.15], the formal discipline of radio astronomy was a direct product of World War II, growing out of the radar tools and techniques developed during the war. After the war, there was a great interest held by former wartime radio engineers in using the recently developed radar and radio communication technologies for scientific purposes, and scientists in the nations that had been involved in the war, especially Australia, Britain, the US, and the USSR, began to pursue research in the newly burgeoning field of radio astronomy [1.16].

Several years after the end of the war, however, as noted by historian Paul Foreman, US physics "underwent a qualitative change in its purposes and character", with increased government intervention and a new emphasis "in the nation's pursuit of security through ever more advanced military technologies" [1.17]. Radio astronomy had a unique dual purpose as both a tool for war and scientific pursuits, which led to heavy investment by governments in the infrastructure and institutions required. This dually motivated investment is perhaps best embodied in the establishment of the US National Radio Quiet Zone. In 1958, the Federal Communications Commission (FCC) had established about 13,000 square miles as the National Radio Quiet Zone (NRQZ), which placed restrictions on radio broadcasting in the area [1.18]. The principal reason given for this large-scale federal intervention was the opening of the first US national observatory for radio astronomy, the National Radio Astronomy Observatory (NRAO), which had been established a couple of years earlier, on November 17, 1956 [1.19]. Buried in the last sentence of the fourth line-item of the FCC's Docket No. 11745, amending the commission's rules and regulations to give interference protection to frequencies utilized for radio astronomy, however, was this statement: "additional coordination would be undertaken by the commission with the Department of Navy at Washington, D.C. with respect to the Sugar Grove facility" [1.20].

The FCC document goes into no further detail on the purpose of the Sugar Grove facility, and the National Reconnaissance Office still to this day keeps much of the documents from the planning and development of facilities in Sugar Grove classified. It is relatively well known from other sources, however, that soon after the establishment of NRAO, the Naval Research Lab began plans to build a 600-foot radio telescope for the purposes of gathering intelligence on the Soviet Union [1.21]. So while the US promoted the "pure science" purpose of the NRQZ, there was an undercurrent of military motivation. The advent of radio astronomy and CETI science during the Cold War is particularly interesting, then, because of the use of its technology for spying on both the unknown intelligence of the cosmos and unknown intelligence here on Earth. As this chapter will demonstrate, watching and being watched were intrinsic parts of collaborative scientific work during the Cold War, and CETI is a particularly revealing example of this fusion of scientific and geopolitical aims.

1.3 Finding the 21-cm Hydrogen Line

Given the influence of the Cold War in developing both radio astronomy and CETI, it is surprising that prior histories of the search for extraterrestrial intelligence have largely neglected contributions from the Soviet Union. This is perhaps because of poor communication between the US and USSR during the Cold War, which meant Soviet radio astronomy had little reach or impact on the rest of the world [1.22]. In his introduction to the English edition of *A Brief History of Radio Astronomy in the USSR*, NRAO astronomer and historian Kenneth Kellermann noted that there were several instances in which Soviets would make a discovery before the rest of the world, but due to communication barriers, Western scientists would often receive credit for making the same discovery at a later date. One example was the case of the discovery of radio recombination lines, which Soviet astronomers at Pulkovo Observatory and Lebedev Physical

Institute (FIAN) [1.23] discovered as early as 1963 or 1964 [1.24]. Because of a lack of clear and consistent communication across borders, however, credit is usually assigned to Bertil Hoglund and Peter Mezger, who reported their own independent discovery in 1965 after observations made at NRAO [1.25]. In addition, while radio astronomy observatories in the US and most other Western countries were civilian organizations, Soviet radio astronomy was inextricably tied to state and military institutions and technologies, and therefore many of their publications were heavily redacted or censored, making it difficult for Western scientists to easily verify their scientific validity [1.26].

Another factor which made attribution of discovery a challenge was the poor exchange of journals and other forms of scientific publishing during this period. The Soviet Union did not join the Universal Copyright Convention, which by 1952 had been adopted by forty nations, including the United States and the vast majority of Western Europe [1.27]. Instead, the Soviet Union subscribed to its own internal copyright law, which granted the right to translate and publish any foreign work without the original author's or publisher's consent, and without the attribution of royalties [1.28]. Additionally, foreign works were also permitted to be edited and modified without the original author's permission, and this was commonly done with journals. The Communist Party control over scientific institutions was "almost absolute", so foreign scientific journals first went through a process of review, translation, and censorship before they were released in the Soviet Union [1.29]. As a result of this ideological stronghold, scientists usually had to expect delays in receiving new scientific results from the West, putting them at a disadvantage to their Western colleagues [1.30]. To make matters of communication worse, Soviet publications were practically inaccessible to Western scientists. As historian of Soviet science Michael Gordin has noted, in the late 1950s only 1.2% of American scientists could read Russian. Furthermore, in 1947 the Soviet Communist Party decided that "publication of Soviet scientific research abroad constituted a betrayal of native scientific resources" and as a result, the leading foreign-language publications were eliminated [1.31]. This lack of consistent sharing of information sometimes led to a difference in attribution of discovery in the West and the Soviet Union, leading to confusion that lasts up to the present day.

One example of this lasting confusing can be seen in historical documentation of the discovery of the 21-cm hydrogen line, the prediction of which is generally generally attributed to Hendrik van de Hulst, a Dutch astronomer and mathematician who published a paper in 1945 that suggested that the transition of neutral hydrogen at 1420 MHz should be theoretically observable using radio telescopes [1.32]. This was an incredibly significant insight in the development of radio astronomy, and later CETI, because of the abundance of hydrogen in the universe. Just a few years later, in 1951, Harvard University astronomers Harold Ewen and Edward M. Purcell became the first to observe the line. Since then, the hydrogen line has become a fundamental aspect of observational radio astronomy, allowing astronomers to map the structure of the Milky Way and other galaxies as well as the large-scale structure of the universe. So significant was its discovery to radio astronomy that it has even been memorialized in song form, with a chorus that reminds listeners that "Ewen and Purcell caught the radiation line/Of interstellar hydrogen, a most important find [1.33]."

When examining the historiography of the discovery of the line, however, another name besides van de Hulst's sometimes appears: Iosif Samuilovich Shklovsky, a Soviet astrophysicist. Several sources, including *The Biographical Encyclopedia of Astronomers* (2007) and Frank Drake's *Is Anyone Out There?* (1992), claim that Shklovsky predicted the existence of the 21-cm hydrogen line entirely independently of van de Hulst. This is a mischaracterization of what occurred, but a somewhat understandable one if unfamiliar with the issue of Soviet journal publishing. Astronomer and historian of astronomy Woody Sullivan, who conducted oral history interviews with Shklovsky before his death in 1985, notes in his book *Cosmic Noise: The History of Early Radio Astronomy* (2009) that Shklovsky was not able to obtain van de Hulst's paper in the Soviet Union because of the aforementioned publication issues. Instead, Shklovsky read a brief mention of van de Hulst's discovery in a review of astronomy published in 1947. The mention of this prediction in the review "lit [him] on fire", and inspired him to attempt to calculate the transition of the line on his own, without access to van de Hulst's original paper [1.34]. In 1949, the same year he received his doctorate, Shklovsky published a paper titled "Monochromatic Radio Emission from the Galaxy and the Possibility of its Observation" [1.35]. Shklovsky credits van de Hulst, with the first line of his Russian-language paper on the subject, asserting: "Van de Hulst was the first to point out the probability of the existence of monochromatic radio emission from the galaxy" [1.36].

Nonetheless, Shklovsky's semi-independent calculations of the 21-cm hydrogen line helped to establish him as a significant early contributor to the development of radio astronomy in the USSR. Yet there is another way the 21-cm line had an impact on Shklovsky's life. Once observed by Ewen and Purcell, the newly important status of this line to radio astronomy inspired two physicists, Philip Morrison and Giuseppe Cocconi, to propose its use in searching for extraterrestrial intelligence. Published in

Nature in 1959, Morrison and Cocconi's paper, "Searching for Interstellar Communications", advocated a search for artificial signals on the hydrogen line, becoming the first scientific publication to propose a CETI observational technique [1.37]. Their rationale in observing at 1420 MHz was that, if an extraterrestrial civilization wanted to send a signal at a frequency that Earth was sure to detect, it would make most sense for them to broadcast it at a frequency that is important to radio astronomy, where humans were already looking. Shklovsky was "greatly impressed" by Cocconi and Morrison's article, which added such great potential purpose to the line he had played a role in discovering [1.38]. Shortly after the publication of the Morrison and Cocconi paper (although conceived before it was published), Frank Drake, an American scientist at the National Radio Astronomy Observatory, conducted what is generally considered the first modern CETI search, using NRAO's 85-1 to search for an extraterrestrial signal near the hydrogen line. His search, named Project Ozma, was conducted in April 1960 and looked for signals from two stars—Tau Ceti and Epsilon Eridani [1.39]. Although the results were non-conclusive, Drake's Project Ozma excited the scientific community, including Shklovsky.

1.4 The Role of the Space Race in Early CETI

The start of the Space Race deepened Shklovsky's fascination with the possibility of searching for extraterrestrial intelligence. In January 1959, shortly before he learned of the Morrison and Cocconi paper, the Soviet spacecraft, *Luna 1*, launched. After the glowing success of the first three *Sputniks*, *Luna 1* became yet another important achievement for the Soviet Union at the start of the Space Race: it was the first spacecraft to reach the vicinity of the moon [1.40]. There was a small dilemma, however: in 1959, the Soviet Union did not yet have a radio telescope capable of tracking its spacecraft launches. It is for this reason that Sir Bernard Lovell, then director of Britain's Jodrell Bank Observatory in Manchester, became a figure of much acclaim in the USSR. His Mark I telescope became the first telescope in the West to track the launch of Sputnik I in 1957, confirming the Soviet achievement to the rest of the world [1.41]. Without their own radio telescope, obtaining the exact coordinates of their rocket launches was a real challenge for the Soviets. For the launch of *Luna 1*, Shklovsky had proposed a solution. With the support of the "chief designer" of the Soviet space program, Sergei Korolev, Shklovsky and his team at Shternburg Astronomical Institute at Moscow State University designed what they called an "artificial comet" to aid optical observations of rocket trajectories [1.42]. This "comet" was composed of a luminescent cloud of sodium gas, which would be ejected from the rocket during launch and photographed, making the spacecraft detectable and therefore easier to track [1.43]. For his work on the *Luna 1* rocket telemetry, Shklovsky was awarded the Lenin Prize, the highest award for scientific achievements bestowed by the Soviet government [1.44].

This success put him on good terms with the president of the Soviet Academy of Sciences, Mstislav Vsevolodovich Keldysh, who was sometimes referred to as the "chief theoretician" of the Soviet space program, in conjunction with Korolev. As a result of his success, Shklovsky was invited to group meetings held in Keldysh's office for "regular [discussions] of space projects" [1.45]. During one such meeting in 1961, Keldysh reminded the group that the five-year anniversary of Sputnik I's launch was only a year away, and "should be properly celebrated". Shklovsky, whose imagination had so been swept up by the Morrison and Cocconi paper and the romance of the burgeoning space age that he felt "like a kid who'd fallen in love", eagerly proposed that he write a popular science book exploring the idea of extraterrestrial life [1.46]. Keldysh approved, but this meant there was only one year to write the book. What might have been a problem to some, Shklovsky saw as an opportunity.

That was because, in addition to the censorship of foreign scientific journal publication in the Soviet Union, there were also censorships placed on internal publications in the Soviet Union. In particular, any works regarding "space" during the Space Race were particularly challenging to get through censors [1.47]. The main organization of press censorship was called the Main Administration for the Protection of Military and State Secrets in the Press under the USSR Council of Ministers—or, "Glavlit" for short, an acronym for its Russian name [1.48]. Glavlit had an associated office called the Commission for Research and Exploitation of Cosmic Space which held specialized censors for space sciences. During the space age, "every book, article, radio or TV broadcast in any way connected with space flights must have an authorization from [that] censorship office" [1.49]. Because of the significance of publishing the book for the fifth Sputnik anniversary, however, Shklovsky suspected he "would have a better chance of escaping the embraces of the general censorship", as censors would have less time to review it [1.50]. Getting past censorship was particularly important for a book on the subject of extraterrestrial life. Although by the 1960s Lysenko's views on biology had begun to fall out of favor, the official end to the ban on criticism of Lysenkoism* was not lifted until the 1980s [1.51]. In writing about the development of life in the universe, Shklovsky intended to "demolish" the theories of Alexander Oparin, Soviet biochemist and "close confederate" of Lysenko, who researched the origins of life on Earth [1.52]. Challenging Oparin was a potentially dangerous undertaking, not to mention challenging. After all, it was difficult to find reputable books on molecular biology in the Soviet Union during the Lysenkoist period [1.53]. Nonetheless, Shklovsky's bet paid off. His book, titled *Universe, Life, Mind* made its way past the censors in its rush to be printed in time for the Sputnik anniversary and it was published in December 1962, despite some "uproar" [1.54].

Despite the initial challenges, *Universe, Life, Mind* became incredibly popular, due in part to the tremendous public impact of Soviet space activities. In the early 1960s, the Soviet Union had been dominating the United States in the so-called "race to space", as it had been labeled by President Kennedy

Figure 1.3 Gagarin waves from the car outside the headquarters of Amalgamated Union of Foundry Workers after receiving a medal, 12th July 1961. From the Daily Herald Archive/National Museum of Science & Media/Science & Society Picture Library.

in 1961 [1.55]. By 1962, the Soviet Union had become the first nation on Earth to launch a satellite into orbit, put the first living creature in space, put the first man in orbit, the first photo of the far side of the moon, launch the first space craft to reach the moon, and send the first probe to impact the moon. Characteristically, the Soviet government capitalized on these victories with great displays of promotional propaganda in the form of films, posters, and parades (Figure 1.3). Iina Kohonen, an expert in space-related visual propaganda and photojournalism in the Soviet Union, has argued that the Soviet space program was composed of a combination of "military aspirations, state propaganda and utopian thinking" [1.56]. It is the latter characteristic, "utopian thinking", which largely fueled Soviet public interest in space, and led to the success of Shklovsky's book, which painted an exciting future of human cosmic exploration. Soviet depictions of the future in space portrayed a universe in which communism had succeeded, the world was peaceful, and humanity dedicated itself to the pursuit of science and exploration of the cosmos. This mentality was not limited exclusively to the Soviet Union. Americans similarly created narratives about a human destiny in space—albeit with a US-centric bent. While Soviet literature espoused spreading a communist ideology throughout the cosmos, Americans viewed space as a "final frontier", an extension of the driving force of manifest destiny [1.57]. Therefore, in the United States, public interest in space exploration was also growing. The US had managed a rocky start in the competition, but especially after Kennedy's exhortation of the goal of a 'moon-shot', the US public imagination was captured by the space age.

1.5 Same Planet, Different Civilizations

Perhaps the most illustrative example of the role the Cold War played in the early development of CETI was Shklovsky's collaboration with a young American astrophysicist, Carl Sagan. While Shklovsky had been designing his artificial comet for the Soviet lunar program in 1959, Sagan was similarly involved in the race to space. At only 25 years old, he had been hired as a consultant to the newly established National Aeronautics and Space Administration (NASA) [1.58]. By August 1962, just a few months before Shklovsky's book was published, Sagan had just seen to the launch of NASA's Mariner 2 mission to Venus, which he had helped to design and manage. Like Shklovsky, Sagan had a fascination with the idea of searching for extraterrestrial intelligence. He was a member of NASA's Planetary Biology Subcommittee and hoped to find signs of life on other planets through his work on NASA missions [1.59].

In April 1961, the year prior to the publication of *Universe, Life, Mind*, Sagan, like Shklovsky, was introduced to the radio search for extraterrestrial intelligence. The newly established National Radio Astronomy Observatory had just carried out the first search for extraterrestrial intelligence using radio telescopes, the aforementioned Project Ozma [1.60]. Although the project was not successful in detecting ETI, it had made a major cultural and scientific impact, inspiring many subsequent observations. As a result of the sudden interest in CETI, the Space Sciences Board (SSB) of the National Academy of Sciences (NAS) decided to support what they called a "quiet meeting" in the subsequent autumn, to discuss the possibility of making radio contact with extraterrestrial intelligence [1.61].

Although small, the first CETI conference was attended by an elite group of scientists from several fields, including NRAO director Otto Struve, future Nobel prize–winning physicist Charles Townes, Nobel prize–winning chemist Melvin Calvin, NASA astrophysicist Su Shu Huang, Project Manhattan scientist Philip Morrison, and John C. Lilly, a biologist who researched dolphin communication [1.62]. The scientists at the conference drew connections between Lilly's work and their own— the pursuit of communication with the non-human and with unknown intelligences. As a result, they dubbed themselves "The Order of the Dolphin", a semi-secret CETI society [1.63]. Sagan was the youngest scientist in attendance, perhaps invited mainly because of his connections with Melvin Calvin, who had written a letter of recommendation that helped him acquire his postdoctoral position at UC Berkeley [1.64].

The SSB outlined three major topics to be addressed at the conference. First, they wanted the attendees to consider the "estimates of limiting values for the probability of existence of planets on which civilized life is likely to have evolved". Second, they should determine whether this number was sufficiently large and decide whether present search methods were useful enough to make the search worth pursuing. And finally, the conference attendees were to make recommendations to the SSB for further study [1.65]. To address this issue of feasibility, Drake devised an equation which was henceforth known as the "Drake Equation". The equation was simple, a product of factors, the values of which were largely unknown at that time. The Drake Equation is represented as "$N = R^* \cdot f_p \cdot n_e \cdot f_l \cdot f_i \cdot f_c \cdot L$", with each variable defined as follows:

> "N = The number of civilizations in the Milky Way galaxy whose electromagnetic emissions are detectable.
> R* = The rate of formation of stars suitable for the development of intelligent life.

f_p = The fraction of those stars with planetary systems.
n_e = The number of planets, per a planetary system associated with one star, with an environment suitable for life.
f_l = The fraction of suitable planets on which life actually appears.
f_i = The fraction of life-bearing planets on which intelligent life emerges.
f_c = The fraction of civilizations that develop a technology that releases detectable signs of their existence into space.
L = The length of time such civilizations release detectable signals into space [1.66]."

Since the equation could not be reasonably calculated with their present knowledge, it was instead meant as a tool that the attendees of the conference could use to frame their estimates of ETI in the galaxy.

The final variable, *L*, has been the focus of many studies in technosignature literature, with estimates ranging from incredibly optimistic to deeply bleak [1.67]. As time has progressed and fields of study such as exoplanetary science have developed, many of the Drake equation variables have begun to fill in, but the answer to *L* has remained elusive and highly relevant. SETI anthropologist Kathryn Denning has argued that *L* is the variable in which "we are exceptionally emotionally invested", because "our estimates of *L* are intertwined with our forecasts for our own civilization's end" [1.68]. For that reason, Drake's inclusion of *L* in the equation is particularly revealing of the cultural climate in which CETI developed—after all, almost exactly one year following the first CETI conference in Green Bank, the world would experience its first nuclear standoff, the Cuban missile crisis. The question of *L* was inextricably tied to concerns about the role of technology in the destruction of life and civilization, concerns which were incredibly pressing during the height of the Cold War.

Shortly after the NRAO CETI conference concluded, Sagan wrote a paper titled "Direct Contact Among Galactic Civilizations by Relativistic Interstellar Spaceflight", in which he made an argument in support of his own estimate for the values in the Drake Equation, including the variable *L*. While he began his paper admitting that the "parameters are poorly known", he nonetheless came to the conclusion that there exists "~106 extant advanced technical civilizations in our Galaxy" [1.69]. Perhaps having heard of his interest in CETI from someone at the conference, Sagan sent a draft of this paper to Shklovsky in 1962, inviting his comments.

In response, Shklovsky asked Sagan if he might be permitted to incorporate Sagan's article into his upcoming book, which was to be published at the end of the year. He ended his letter to Sagan with a joke—wondering if perhaps they might someday meet, although remarking "the probability of this event is no smaller than the probability of a visitation of the earth by astronauts" [1.70]. Sagan agreed to have his paper incorporated into *Universe, Life, Mind*, and in return asked Shklovsky if he had any plans to publish an English translation of the book in the United States. If not, Sagan proposed allowing him to aid with such a publication himself [1.71]. Shklovsky agreed to the collaboration, but his prediction that they would work together without meeting was correct. The pair would not meet in person until well after the publication of the English book. This is largely because of constraints put upon Shklovsky because of his standing in the eyes of the Soviet authorities—yet another example of barriers to communication placed upon scientists during the Cold War period.

Although Shklovsky was held in high regard by his colleagues for his work early in the Space Race, his outspoken personality prevented him from experiencing the mobility and access granted to some of his peers. His letter to Sagan regarding the unlikelihood of their meeting reflected the personal resentment Shklovsky held towards Soviet bureaucracy for limiting his ability to collaborate with international peers. In a short autobiographical story published posthumously, Shklovsky recounted his first expedition outside the Soviet Union; a research trip to Brazil in 1947, to observe a solar eclipse. Shklovsky wrote of the time:

> I took it for granted that the forthcoming expedition to the Tropic of Capricorn, to a faraway Brazil as beautiful as anything in a fairytale, was just the beginning and that many more fine and soul-stirring things yet unknown lay ahead. After a poverty-stricken youth and the harsh suffering of the war years, the world had at last opened up for me [1.72].

Unbeknownst to him at the time, Shklovsky would not be allowed to travel abroad for another nineteen years. Despite his many accomplishments, including the aforementioned Lenin Prize for his work during the Space Race, Shklovsky often faced travel bans, and was never elected as a full member of the Soviet Academy of Sciences, which he resentfully concluded was because of his Jewish heritage and commitment to promoting human rights [1.73].

In this collection of autobiographical stories, humorously titled *Five Billion Vodka Bottles to the Moon* (1991), Shklovsky also recounted his memory of the "astronomers purge" during Stalin's Great Terror of 1936-38, during which over two dozen leading Soviet astronomers were arrested, many of whom were later executed, or died in the Gulag [1.74]. The purge occurred in part because of an astronomy PhD student who failed his candidacy exam, and afterwards wrote a letter of denunciation of the astronomer who administered his examination, Boris Numerov [1.75]. Denunciations were common practice in the Soviet Union, most especially during the Stalinist era, and according to historian Sheila Fitzpatrick, largely fell into three categories of accusation: political disloyalty, "concealment of class identity", and "abuse of power" [1.76]. The state would investigate the accused, and punish those who they deemed guilty. This, however, led to the problem of individuals abusing the system by using the state to "settle personal scores or advance the denouncer's individual interests", as was clearly the case with the failed PhD student [1.77]. In his denunciation, he accused his examiner of having "foreign contacts", and shortly afterward the NKVD began investigating Numerov. After arresting and torturing him, he confessed "to being the organizer of a counterrevolutionary group of astronomers and geophysicists that had cooperated with German fascists and had engaged in wrecking, spying, and terror since 1929" [1.78]. In his confession, Numerov listed nearly the entire astronomy community as co-conspirators, setting off a chain of subsequent arrests and denunciations which historian and foreign diplomat Robin Mccutcheon argued led to the "disappearance" of approximately "10 percent to 20 percent of all astronomers in the Soviet Union in 1935" [1.79]. Only a graduate student at the time of the purge, Shklovsky avoided being arrested, but after his death, his account of the events leading up to the purge was published, a clear indication that the memory of the ordeal remained with him for the rest of his life [1.80]. Still, despite the threat of denunciation, Shklovsky often risked life and freedom by standing for what he believed in, and continually criticized the USSR for, among other things, its anti-Semitism and restriction of scientific freedom.

Even as his fellow physicists occasionally disappeared or died under mysterious circumstances, Shklovsky would defend mistreated colleagues and condemn what he viewed as ethical transgressions committed by the Academy or Soviet government. For example, in 1973, approximately forty members of the Soviet Academy of Sciences signed a document condemning Soviet nuclear physicist and political dissident Andrei Sakharov in a denunciation [1.81]. Shklovsky, on the other hand, sent in a strong letter of advocacy for Sakharov. As a result of this daring move, he was forbidden to appear at international scientific meetings he had previously been allowed to attend, such as the International Astronomical Union meetings. When asked by Western colleagues about Shklovsky's absence from international conferences, Soviet officials would reply "his health is too poor" to travel abroad [1.82]. An American colleague encountered Shklovsky in the USSR during this period of travel ban and enquired about his health, to which Shklovsky wryly replied, "Yes, I have diabetes. Too much sakharov" [1.83]. This was a joke because the word for "sugar" in Russian is "sakhar".

Although he avoided arrest during the Stalinist era, Shklovsky's dissidence sometimes still put his life in danger. One register of this was relayed much later by Sir Bernard Lovell, the aforementioned British physicist and radio astronomer who was involved in British tracking of Soviet satellites as an anti-nuclear tactic during the Cold War. Lovell, in his capacity as the director of the Jodrell Bank Observatory, went on an "unprecedented" scientific visit to the Deep-Space Communication Centre in the USSR in 1963, the year after *Universe, Life, Mind* was published [1.84]. The official reason for visit was because of a planned international scientific collaboration; the Jodrell Bank Observatory was to assist in upcoming Soviet-US communication satellite experiments [1.85]. During his stay in the USSR, however, Lovell learned that the Soviets were considering abandoning their moonshot, and his report upon return to the West played a key role in Kennedy's consideration of a joint Soviet-US moonshot [1.86]. Also during this visit, the Soviets attempted to convince Lovell to stay in the USSR and join the Soviet Academy of Sciences, perhaps because, as mentioned earlier in the chapter, Lovell was held in extremely high regard in the Soviet Union due to his role in tracking early Soviet satellites and probes, validating Soviet success in the West [1.87]. Lovell's refusal of their offer, alongside his newfound insight into future Soviet moon plans, may have led to the development of a dangerous situation. According to Lovell, Soviet colleagues attempted to assassinate him using a lethal dose of radiation while on his tour [1.88].

During his life, Lovell never spoke in detail about the alleged assassination attempt. He did, however, write an account of this event and gave it to the John Rylands Archive at the University of Manchester, where it was to remain unpublished until after he died, in 2012. It is unclear whether the events Lovell described truly occurred as he perceived them—it may well have been a product of Lovell's paranoia and the many myths and rumors enveloping the secretive state. Nonetheless, it is clear tensions were high, and the threat of danger was present, at the very least, in the minds of the scientists from both countries. Lovell's memorandum also detailed an encounter with an unnamed Soviet scientist who implored him to invite Shklovsky to his laboratory in England, for Shklovsky's life was "in great danger" [1.89]. The nature of the danger and the reason for it remained a mystery to Lovell, and there are no available sources that might explicitly explain the cause. It may have been the result of Shklovsky's open dissent to the treatment of Jewish academics, which likely cost him the ability to travel to the US between 1947 and 1967, or perhaps it was the biology controversy stirred up from his recent publication of *Universe, Life, Mind.* Either way, as evidenced by the tense Soviet political atmosphere, the fear the unnamed Soviet astronomer held for Shklovsky's safety was not entirely unfounded. Nevertheless, despite this danger, Shklovsky was able to make and maintain relationships with scientists around the world by extensive correspondence, including with Carl Sagan.

1.6 Making *Intelligent Life in the Universe*

Yet establishing these international correspondences did not come easily, and the road to the publication of the English translation of Shklovsky's book was rife with difficulties. When the US publishers, Holden-Day, wrote Shklovsky in January 1963 to confirm their agreement to publish the translation, they stated that they intended to "make every effort to publish the book in the shortest possible time and [estimated] that this could be done in three to four months..." [1.90]. Over eight months later, however, Shklovsky had no update from Sagan or the publishers. Feeling frustrated that he had not heard from Sagan within the few months the publishers had assured, an irate Shklovsky wrote to Sagan asking, "do you deny me the courtesy of informing of the status of the American translation of the book?" [1.91] Unbeknownst to him, however, Sagan had sent him several letters regarding the translation and its progress. Shklovsky had simply not received them—an example of yet another type of crisis in Earth-bound communication common during the Cold War: the unreliability of postal correspondence across the Iron Curtain.

The issue of Soviet interference with the post is well documented. In the 1980s, for example, there was a concerted effort by the US government to address the issue of deliberate interference with the flow of mail between the US and USSR. A 1989 report on the history of mail interruption prepared for the US Committee on Post Office and Civil Service of the House of Representatives examined what they described as a "long existing problem" [1.92]. The report claimed that over many years, a "significant number" of items of communication sent from the US to USSR had "disappeared, or were opened, inspected, and/or confiscated by officials of the Soviet Union, without the proper notification given to mailers" [1.93]. Mail interruption was described as a "violation of human rights" by the authors of the report, who also presented data which indicated the rate at which letters were not delivered. For example, in 1985, non-delivery of letters was at approximately 12 percent [1.94]. Given that 1985 was the year *perestroika* began, it is reasonable to assume earlier decades might have faced higher rates of mail interruption, although data to support this hypothesis is difficult to attain.

It would be naïve, of course, to assume the problem of mail disruption existed exclusively on the side of the Soviets. As Edward Pessen, historian of American history, has pointed out, "what a nation's leaders call its policy is after all only its stated policy" [1.95]. Although United States policy was purported to be emblematic of a "free" society, there is also evidence of US interference with postal communication. As Pessen points out, "in blatant violation of the law creating it, the CIA kept files and spied on American citizens, tampered with and opened the mail of hundreds of thousands" [1.96].

What differed between the US and USSR, however, were the populations targeted by mail disruption. For example, both the authors of the 1989 report, as well as an earlier congressional hearing on disruption of mail service in 1984, argued that Soviet Jewish people were disproportionately impacted by mail disruption [1.97]. The report argued that these obstacles were indicative of an effort to "isolate the Jewish community in general, and the 'activists' in particular, from any contact with the outside world" [1.98]. The Soviet Union practiced anti-Zionist campaigns throughout the latter half of the 20th century, fueled in part by its negative relations with Israel during the Cold War. Anti-Zionist propaganda (Figure 1.4) promoted historically anti-Semitic representations of Jewish people, which in turn provoked anti-Semitic attitudes within the Soviet Union, leading to negative impacts on Jewish Soviets [1.99]. In a state where denunciations on political or social loyalty could lead to deadly consequences, questioning Jewish loyalty through anti-Zionist campaigns was no small accusation. As both a Jewish man and someone who might have been perceived as an activist, it is not unreasonable to assume anti-Semitism might have been a reason for Shklovsky's difficulties with international correspondence. Perhaps it was in part this discrimination that later led Sagan to reflect, in his science fiction novel *Contact* (1986), that human beings may have difficulty effectively communicating with extraterrestrial intelligence so long as they continue to create aliens among themselves through the process of othering and oppression [1.100].

During the publication process, further challenges had been presented regarding the illustrations in *Universe, Life, Mind*. As noted earlier in this chapter, the Soviet Union did not join the Universal Copyright Convention, and therefore did not need to receive permissions from creators outside the Soviet Union to reproduce them in print. The United States was a member

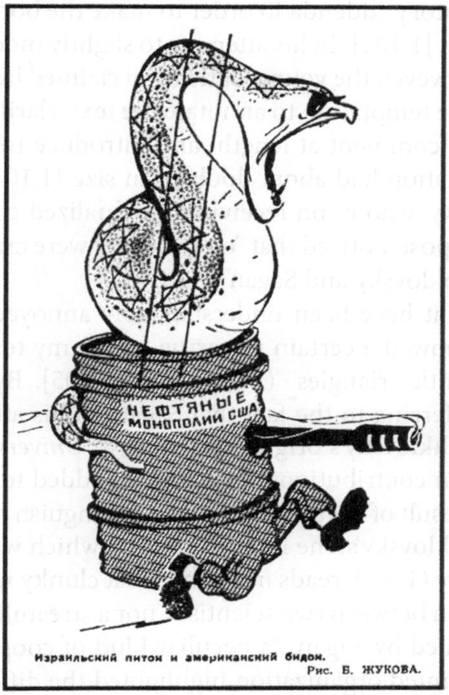

Figure 1.4 'The Israeli python and the American barrel,' B. Zhukov, Pravda Vostoka, Feb. 11, 1968. (From The Israeli-Arab Conflict in Soviet Caricatures, 1967–1973 by Yeshayahu Nir, Tcherikover Publishers, 1976).

of that convention, however, and therefore it was a hefty task tracking down and gaining permissions to reproduce the illustrations in the US. Although this caused delay, it was not the main reason the book was not yet ready for publication.

In an earlier letter to Sagan, under the mistaken impression that he was a trained biologist, Shklovsky commented that Sagan was welcome to make changes and additions to the biological and genetic sections of the book, adding that he believed he would be much more competent in these matters, especially since Shklovsky had limited access to biology texts due to Lysenkoism [1.101]. Decades later, Shklovsky claimed that Sagan "interpreted my request broadly", which, given what happened, might be considered an understatement [1.101]. In the introduction to Intelligent Life in the Universe, Sagan explained the delay, stating that he felt it was necessary to include explanatory addenda in order to make the book more accessible to Western readers [1.102]. In his attempts to slightly modify the book for a US readership, however, the young Carl Sagan claimed he "found [himself] unable to resist the temptation to annotate the text, clarify concepts for the scientific layman, comment at length, and introduce new material", until the English translation had about doubled in size [1.103]. This came as a shock to Shklovsky, who upon receiving the finalized copy of the English translation in the post, noticed that "on the cover were crammed the names of two authors: Shklovsky and Sagan" [1.104].

Shklovsky might have been understandably annoyed by this but said of Sagan: "[he] showed a certain integrity; he left my text unchanged and set off his with little triangles" (Figure 1.5) [1.105]. By "little triangles", Shklovsky was referring to the fact that Sagan, in his attempts to be clear about what was Shklovsky's original wording in *Universe, Life, Mind*, and what were his own contributions, enclosed his added text within inverted delta signs. As a result of Sagan's attempts to distinguish his own words and thoughts from Shklovsky's, the English edition, which was titled *Intelligent Life in the Universe* (1966), reads in a somewhat clunky manner. It was neither a conversation between two scientists, nor a streamlined narrative, but instead, as described by Sagan, "a peculiar kind of cooperative endeavor" [1.106]. The disjointed organization highlighted the different perspectives of the two scientists, which served an extremely valuable purpose. On the one hand, it gave readers an insight into the different ways a Soviet and

▽ The appearance of a supernova in the spiral galaxy NGC 4725 may be seen in Figure 7–1. The top photograph was taken on May 10, 1940, when the supernova exceeded in brightness all other regions of the spiral arms of this galaxy, except for the galactic nucleus; the bottom photograph was taken on January 2, 1941, when no supernova was evident. Figure 7–2 shows a less spectacular but more common variety of supernova; this occurred in the nearer galaxy M101. Here too, we have a "before" and "after" sequence, with the arrow indicating the supernova △

Despite the infrequency of supernovae in the Milky Way, a number have been recorded in historical times. On July 4 ▽ sic! △, 1054, a "guest-star" appeared in the sky; it was duly reported by Chinese scholars. This star was so bright that it could be seen during the daylight hours. It surpassed Venus in luminosity; only the Moon and Sun were brighter. For several months this star was visible to the naked eye; then it gradually faded from view.

Figure 1.5 Excerpt from *Intelligent Life in the Universe* (1966). Sagan made contributions to Shklovsky's original text by interjecting between "∇Δ" symbols. In the second paragraph above, Sagan inserts a tongue-in-cheek 'sic erat scriptum' to poke fun at the unintentional reference to the American Day of Independence.

an American astrophysicist approached the problem of extraterrestrial communication. But additionally, and perhaps most crucially, it prevented Shklovsky from facing problems in the Soviet Union for what it would have viewed as undesirable arguments deployed within the book. Shklovsky was able to get the first edition of *Universe, Life, Mind* past censors because of the fortuitous circumstances outlined earlier, but the new edition, with all of Sagan's cultural additions, would face greater problems. Sagan pointed out in the introduction to the book:

> As the reader might expect for a book written by two authors, one in the Soviet Union and one in the United States, there are occasional ideological differences. I have not tried to avoid these problems, but I also have not tried, in what is primarily a scientific work, to rebut each ideological assertion. When Shklovsky expresses his belief that lasting world peace is impossible while capitalism survives, or implies that lasers are being developed in the United States for their possible military applications alone, I have let the content of these statements stand, despite their political intent [1.107].

Shklovsky later noted the benefit of Sagan's approach to distinguishing their words, later claiming he realized "that my American 'coauthor' had done me a priceless boon in distinguishing his text with triangles. Otherwise our vigilant official "readers" could have made things tough for me" [1.108]. Clearly, the challenges faced by Shklovsky and Sagan in their attempts to work with each other across the Iron Curtain demonstrates that the difficulty of communicating on Earth posed nearly as large a challenge as communicating with extraterrestrials.

1.7 Conclusion: Returning to L

Intelligent Life in the Universe also addressed the *L* variable from the Drake Equation that had been introduced to Sagan at the 1961 Green Bank conference. As noted earlier in the chapter, *L* referred to the longevity of any given intelligent extraterrestrial civilization. The inclusion of this variable in the Drake Equation was telling for both the period and the culture in which it was presented. In 1961, when the equation was presented, Cold War tensions were reaching their height. In the 1960s and 1970s, the United States alone possessed 1,054 nuclear missiles [1.109], over ten times more than necessary to make the Earth hostile to human life [1.110]. For the first time in human civilization, human beings had the capacity to destroy their entire species in a nuclear war. This realization led Sagan and Shklovsky to consider the following in *Intelligent Life in the Universe*:

> Another question of some relevance to our own time, and one whose interest is not restricted to the scientists alone, is this: Do technical civilizations tend to destroy themselves shortly after they become capable of interstellar radio communications? [1.111]

This question, posed by a Soviet and American, is demonstrative that CETI's reach extended far beyond physics and cosmology, and promoted questions of a philosophical, historical, and sociological bent. In 1971, a few years after the publication of *Intelligent Life in the Universe*, Sagan and Shklovsky held a joint US-USSR conference on the topic of communication with extraterrestrial intelligence [1.112]. Among those invited was a historian, William McNeil, who had achieved notoriety for his earlier publication of *Rise of the West* (1963), which argued that contact and communication between foreign civilizations was the primary driver of human history. The presence of a historian whose expertise was in the rise and fall of civilizations reflected the pressing concern L held in the minds of CETI researchers. This preoccupation with the destruction of civilizations led Sagan, less constrained than Shklovsky in speaking out against nuclear weaponry (which in part had led the aforementioned Sakharov to exile), to dedicate much of his effort to anti-nuclear activism. In 1983 he published an essay titled "Nuclear War and Climactic Catastrophe: Some Policy Implications" in *Foreign Affairs*, where he argued that if the US and USSR did not reverse their arms race, "there is a real danger of the extinction of humanity" [1.113]. The aforementioned Philip Morrison, co-author of the paper which had inspired Shklovsky to pursue CETI, had served as a leader on the Manhattan Project and oversaw the assembly of the bomb which would detonate above the city of Nagasaki. After viewing the devastation in Japan as part of the Manhattan Project's survey team, Morrison became an adamant anti-nuclear activist and founded the Federation of American Scientists and the Institute for Defense and Disarmament Studies. In addition to his activism, he dedicated much of his life after the war to CETI and chaired the early NASA SETI workshops and studies [1.114]. At least in part, the search for intelligent life on other worlds had prompted its practitioners to fight to preserve civilization on Earth.

It will be surprising to some, in opening a chapter of a book on technosignature research, to find that this chapter did not concern extraterrestrials. Rather, this chapter demonstrates that the history of CETI largely concerns humans and their beliefs, cultures, and biases. As a science, it was also inextricably tied up in the Cold War, posing many communication challenges to the scientists involved, but also forcing them to confront larger philosophical questions. In some sense, the search for technosignatures as we know it today could not have existed without the enormous investment by US and Soviet governments into major scientific projects, such as the establishment of NRAO, the funding of conferences, or the support for the publication of books. Similarly, without the Space Race, there might have been far less public support and individual interest in pursuing the CETI problem. Furthermore, it was the Cold War mindset that influenced CETI scientists such as Frank Drake to pose the question of longevity of civilizations, and that led others, such as Sagan, into anti-war activism. Although the Cold War played a role in the establishment of CETI, it also presented many challenges to those who pursued it and highlighted the cultural and communication barriers between scientists in the US and USSR. In striving to overcome interterrestrial cultural and communications difficulties, and in recognizing the existential dilemma posed by the Cold War, CETI scientists in some sense engaged far more in philosophical and historical problems than they did technical ones. Understanding the historical context in which CETI developed is fundamental to the continued pursuit of the search for technosignatures and communication with extraterrestrial intelligence, as scientists to this day continue to pose philosophical questions which are influenced by our present cultural and geopolitical circumstances.

References

[1.1] A note about terminology: Historically, the search for technosignatures was called CETI (communication with extraterrestrial intelligence) in the 1960s and early 70s. It slowly transitioned to SETI (search for extraterrestrial intelligence) shortly after a US-USSR conference in 1971. Today, the term "technosignatures" is becoming favored, as a compliment to the term "biosignatures" in exoplanetary research. For historical accuracy, this chapter will predominantly use "CETI" when discussing work conducted in the 1960s.

[1.2] Michaud, M. A. G. (2007) Contact with Alien Civilizations: Our Hopes and Fears about Encountering Extraterrestrials. Copernicus Books, New York: 292–293.
[1.3] Dick, S. J. (1982) Plurality of Worlds: The Origins of the Extraterrestrial Life Debate from Democritus to Kant. Cambridge University Press, Cambridge.
[1.4] Huygens, C. (1698) ΚΟΣΜΟΘΕΩΡΟΣ (English translation of Latin: Cosmotheoros: The celestial worlds discover'd: or, Conjectures concerning the inhabitants, plants and productions of the worlds in the planets). Timothy Childe, London.
[1.5] Lowell, P. (1906) Mars and its Canals. Macmillan Company, London: 173.
[1.6] Lowell, P. (1907) A General Method for Evaluating the Surface-Temperature of the Planets; with Special Reference to the Temperature of Mars. Philosophical Magazine 6(1): 161-176.
[1.7] Wallace, A.R. (1907) Is Mars Habitable?: A Critical Examination of Professor Percival Lowell's Book "Mars and Its Canals". Macmillan, London: 110.
[1.8] Dick, S.J. (1998) Life on Other Worlds: The 20th-Century Extraterrestrial Life Debate. Cambridge University Press, Cambridge: 201.
[1.9] "Radio to Stars, Marconi's Hope", The New York Times, January 20, 1919.
[1.10] "Radio to Stars, Marconi's Hope", The New York Times, January 20, 1919.
[1.11] "Radio to Stars, Marconi's Hope", The New York Times, January 20, 1919.
[1.12] "Marconi Testing His Mars Signals", The New York Times, January 29, 1920.
[1.13] "Marconi Testing His Mars Signals", The New York Times, January 29, 1920.
[1.14] "Hello Earth! Hello!", The Tomahawk., March 18, 1920, Image 6. Digital scan held in the Library of Congress. http://chroniclingamerica.loc.gov/lccn/sn89064695/1920-03-18/ed-1/seq-6/ (Accessed 5 November 2017).
[1.15] Jansky, K.G. (1933). Radio waves from outside the solar system. Nature. 132(3323): 66.
[1.16] Sullivan III, W. (2009) Cosmic Noise: A History of Early Radio Astronomy. Cambridge University Press, Cambridge.
[1.17] Forman, P. (1987) Behind Quantum Electronics: National Security as Basis for Physical Research in the United States, 1940–1960, Historical Studies in the Physical Sciences 18(1): 150.
[1.18] Balser, D.S., Ghigo, F.D., and Lockman, F.J. (2016) But It Was Fun. National Radio Astronomy Observatory, Charlottesville: xi.
[1.19] B Balser, D.S., Ghigo, F.D., and Lockman, F.J. (2016) But It Was Fun. National Radio Astronomy Observatory, Charlottesville: xi.
[1.20] FCC Docket No. 11745, National Radio Astronomy Observatory Archives.
[1.21] van Keuren, D.K. (2001) Cold War Science in Black and White: US Intelligence Gathering and Its Scientific Cover at the Naval Research Laboratory, 1948-62. Social Studies of Science 31(2): 207–29.

[1.22] Kellermann, K.I. (2012) Introduction. In: A Brief History of Radio Astronomy in the USSR. S. Y. Braude, A. E. Salomonovich, V. A. Samanian, I. S. Shklovsky, R. L. Sorochenko, V. S. Troitskii, K. I. Kellermann, B. A. Dubinskii, N. L. Kaidanovskii, N. S. Kardashev, M. M. Kobrin, A. D. Kuzmin, A. P. Molchanov, Yu. N. Pariiskii, O. N. Rzhiga, (eds.) Springer, New York: vi.

[1.23] Lebedev's acronym does not fit its English title, but rather its Russian name, which is Физи́ческий институ́т имени П.Н.Лебедева Российской академии наук (Fizicheskii Institut Imeni P.H.Lebedeva Rossiiskoi Akademii Nauk) – FIAN.

[1.24] Kellermann, K.I. (2012) Introduction. In: A Brief History of Radio Astronomy in the USSR. S. Y. Braude, A. E. Salomonovich, V. A. Samanian, I. S. Shklovsky, R. L. Sorochenko, V. S. Troitskii, K. I. Kellermann, B. A. Dubinskii, N. L. Kaidanovskii, N. S. Kardashev, M. M. Kobrin, A. D. Kuzmin, A. P. Molchanov, Yu. N. Pariiskii, O. N. Rzhiga, (eds.) Springer, New York: 40.

[1.25] Kellermann, K.I. (2012) Introduction. In: A Brief History of Radio Astronomy in the USSR. S. Y. Braude, A. E. Salomonovich, V. A. Samanian, I. S. Shklovsky, R. L. Sorochenko, V. S. Troitskii, K. I. Kellermann, B. A. Dubinskii, N. L. Kaidanovskii, N. S. Kardashev, M. M. Kobrin, A. D. Kuzmin, A. P. Molchanov, Yu. N. Pariiskii, O. N. Rzhiga, (eds.) Springer, New York: vi.

[1.26] Kellermann, K.I. (2012) Introduction. In: A Brief History of Radio Astronomy in the USSR. S. Y. Braude, A. E. Salomonovich, V. A. Samanian, I. S. Shklovsky, R. L. Sorochenko, V. S. Troitskii, K. I. Kellermann, B. A. Dubinskii, N. L. Kaidanovskii, N. S. Kardashev, M. M. Kobrin, A. D. Kuzmin, A. P. Molchanov, Yu. N. Pariiskii, O. N. Rzhiga, (eds.) Springer, New York: v.

[1.27] Universal Copyright Convention, with Appendix Declaration relating to Articles XVII and Resolution concerning Article XI (1952). UNESCO. http://portal.unesco.org/en/ev.php-URL_ID=15381&URL_DO=DO_TOPIC&URL_SECTION=201.html.

[1.28] Cramer, A.P. (1965) International Copyright and the Soviet Union. Duke Law Journal: 532.

[1.29] Graham, L. (1972) Science and Philosophy in the Soviet Union. Knopf, New York: 20.

[1.30] Graham, L. (1972) Science and Philosophy in the Soviet Union. Knopf, New York: 20.

[1.31] Gordin, M.D. (2016) The Dostoevsky Machine in Georgetown: scientific translation in the Cold War. Annals of Science, 73(2): 208-223.

[1.32] van de Hulst, H. (1945) The Origin of Radio Waves from Space. Nederlandsch Tijdschrift voor Natuurkunde 11(210).

[1.33] "Hydrogen Line", lyrics © 1995 by Dr. H. Paul Shuch, https://www.qsl.net/n6tx/poetry/setisong/hydrogen.htm.

[1.34] Sullivan III, W. (2009) Cosmic Noise: A History of Early Radio Astronomy. Cambridge University Press, Cambridge: 397.
[1.35] Shklovsky, I.S. (1949) Monochromatic Radio Emission from the Galaxy and the Possibility of its Observation. Астрономический журнал 26: 10-14.
[1.36] Shklovsky, I.S. (1949) Monochromatic Radio Emission from the Galaxy and the Possibility of its Observation. Астрономический журнал, 26: 10.
[1.37] Morrison, P. and Cocconi, G. (1959) Searching for Interstellar Communication. Nature 184(4690): 844-846.
[1.38] Shklovsky, I.S. (1991) Five Billion Vodka Bottles to the Moon. WW Norton & Co. Inc., New York: 250.
[1.39] 85-1 Telescope Log Book, The National Radio Astronomy Observatory Archives.
[1.40] Luna 1 (2020). NASA Space Science Data Coordinated Archive. https://nssdc.gsfc.nasa.gov/nmc/spacecraft/display.action?id=1959-012A.
[1.41] Grahn, S. (2008) Jodrell Bank's role in early space tracking activities - Part 1. Jodrell Bank Centre for Astrophysics. http://www.jb.man.ac.uk/history/tracking/part1.html.
[1.42] Shklovsky, I.S. (1991) Five Billion Vodka Bottles to the Moon. WW Norton & Co. Inc., New York: 248.
[1.43] Shklovsky, I. S., Esipov, V. F., Kurt, V. G., Moroz, V. I., & Shcheglov, P. V. (1959) An Artificial Comet. Soviet Astronomy 36(6): 986.
[1.44] Kardashev, N.S. and Marochnik, L.S. (2017) The Shklovsky Phenomemon. Astronomical and Astrophysical Transactions 30(2): 119–124.
[1.45] Shklovsky, I.S. (1991) Five Billion Vodka Bottles to the Moon. WW Norton & Co. Inc., New York: 248.
[1.46] Shklovsky, I.S. (1991) Five Billion Vodka Bottles to the Moon. WW Norton & Co. Inc., New York: 247.
[1.47] Oral history interview with Leonid Gurvits, conducted by Rebecca Charbonneau at AstroSpace Centre, Moscow, Russia, October 2019.
[1.48] Vladimirov, L. (1971) Glavlit: How the Soviet Censor Works. Index on Censorship 1(3-4): 31.
[1.49] Vladimirov, L. (1971) Glavlit: How the Soviet Censor Works. Index on Censorship 1(3-4): 35.
[1.50] Shklovsky, I.S. (1991) Five Billion Vodka Bottles to the Moon. WW Norton & Co. Inc., New York: 250. Lysenkoism refers to the state-sanctioned dogmatic adherence to the pseudo-scientific research of Soviet agronomist Trofim Denisovich Lysenko in the 1940-50s, during which Lysenko's neo-Lamarkian belief in acquired characteristics as applied to agriculture practices led to mass famines. For more, see: Kean, Sam (2017) The Soviet Era's Deadliest Scientist Is Regaining Popularity in Russia. The Atlantic. https://www.theatlantic.com/science/archive/2017/12/trofim-lysenko-soviet-union-russia/548786/.

[1.51] Gordin, M.D. (2018) Lysenko Unemployed: Soviet Genetics after the Aftermath. Isis 109(1): 73.
[1.52] Shklovsky, I.S. (1991) Five Billion Vodka Bottles to the Moon. WW Norton & Co. Inc., New York: 250.
[1.53] Shklovsky, I.S. (1991) Five Billion Vodka Bottles to the Moon. WW Norton & Co. Inc., New York: 250.
[1.54] Shklovsky, I.S. (1991) Five Billion Vodka Bottles to the Moon. WW Norton & Co. Inc., New York: 251.
[1.55] John F. Kennedy Moon Speech - Rice Stadium. Johnson Space Centre. https://er.jsc.nasa.gov/seh/ricetalk.htm.
[1.56] Kohonen, I. (2009) The Space Race and Soviet Utopian Thinking. The Sociological Review 57: 115.
[1.57] Berry Buzan (2010) America in Space: The International Relations of Star Trek and Battlestar Galactica. Millennium: Journal of International Studies 39(1): 175-180.
[1.58] Carl Sagan (1934-1996). NASA Science Solar System Exploration. https://solarsystem.nasa.gov/people/660/carl-sagan-1934-1996/.
[1.59] Morowitz, H. (1967) Life on Venus. Nature 215: 1259-1260.
[1.60] Project Ozma press release, 1961 SETI Conference, SETI Files, National Radio Astronomy Observatory Archives.
[1.61] Letter from Hugh Odishaw to Otto Struve July 1961, SETI Files, National Radio Astronomy Observatory Archives.
[1.62] List of attendees, 1961 SETI Conference, SETI Files, National Radio Astronomy Observatory Archives.
[1.63] Order of the Dolphin members list, 1961 SETI conference, SETI Files, National Radio Astronomy Observatory Archives.
[1.64] Davidson, Keay (1999) California Cosmology: Carl Sagan: The Berkeley Years. SF Gate. https://www.sfgate.com/magazine/article/CALIFORNIA-COSMOLOGY-Carl-Sagan-The-Berkeley-3490527.php
[1.65] Letter from Hugh Odishaw to Otto Struve July 1961, 1961 SETI conference, SETI Files, National Radio Astronomy Obervatory Archives.
[1.66] Drake Equation. The SETI Institute. https://www.seti.org/drake-equation-index.
[1.67] Shostak, S. (2012) The Value of 'L' and the Cosmic Bottleneck in Cosmos and Culture: Cultural Evolution in a Cosmic Context, Dick S.J. and Lupisella M. (eds). National Aeronautics and Space Administration Government Printing Office, Washington, D.C.
[1.68] Denning, K. (2011) 'L' on Earth. In: Civilizations Beyond Earth: Extraterrestrial Life and Society. Vakoch D.A. and Harrison A.A. (eds). Berghahn Books, New York: 74.
[1.69] Sagan, C. (1963) Direct Contact Among Galactic Civilizations by Relativistic Interstellar Spaceflight. Planetary and Space Science, 11: 485-498.
[1.70] Drake, Frank and Sobel, Dava (1992) Is Anyone Out There? The Scientific Search for Extraterrestrial Intelligence. Delacorte Press, New York: 100.

[1.71] Drake, Frank and Sobel, Dava (1992) Is Anyone Out There? The Scientific Search for Extraterrestrial Intelligence. Delacorte Press, New York: 100.
[1.72] Shklovsky, I.S. (1991) Five Billion Vodka Bottles to the Moon. WW Norton & Co. Inc., New York: 9.
[1.73] Friedman, H. (1991) Introduction. In: I.S. Shklovsky, Five Billion Vodka Bottles to the Moon, 17.
[1.74] Mccutcheon, R.A. (1991) The 1936-1937 Purge of Soviet Astronomers. Slavic Review 50(1): 100.
[1.75] Mccutcheon, R.A. (1991) The 1936-1937 Purge of Soviet Astronomers. Slavic Review 50(1): 108.
[1.76] Fitzpatrick, S. (1994) The Practice of Denunciation in Stalinist Russia. The National Council for Soviet and East European Research, Washington, D.C.: III.
[1.77] Fitzpatrick, S. (1994) The Practice of Denunciation in Stalinist Russia. The National Council for Soviet and East European Research, Washington, D.C.: III.
[1.78] Mccutcheon, R.A. (1991) The 1936-1937 Purge of Soviet Astronomers. Slavic Review 50(1): 108-111.
[1.79] Mccutcheon, R.A. (1991) The 1936-1937 Purge of Soviet Astronomers. Slavic Review 50(1): 108-111.
[1.80] Shklovsky, I.S. (1988) Nevydumannye rasskazy. Eniergiia (6): 41-42.
[1.81] Shklovsky, I.S. (1991) Five Billion Vodka Bottles to the Moon. WW Norton & Co. Inc., New York: 17.
[1.82] Shklovsky, I.S. (1991) Five Billion Vodka Bottles to the Moon. WW Norton & Co. Inc., New York: 17.
[1.83] Shklovsky, I.S. (1991) Five Billion Vodka Bottles to the Moon. WW Norton & Co. Inc., New York: 17.
[1.84] The Kennedy Proposal for a Joint Moon Flight. NASA History Office. https://history.nasa.gov/SP-4209/ch2-4.htm.
[1.85] The Kennedy Proposal for a Joint Moon Flight. NASA History Office. https://history.nasa.gov/SP-4209/ch2-4.htm.
[1.86] The Kennedy Proposal for a Joint Moon Flight. NASA History Office. https://history.nasa.gov/SP-4209/ch2-4.htm.
[1.87] University of Manchester, John Rylands Archive, Memorandum by Sir Bernard Lovell on the Files Covering Contact with Soviet Scientists and Visits to the Soviet Union, 1963 File.
[1.88] University of Manchester, John Rylands Archive, Memorandum by Sir Bernard Lovell on the Files Covering Contact with Soviet Scientists and Visits to the Soviet Union, 1963 File.
[1.89] University of Manchester, John Rylands Archive, Memorandum by Sir Bernard Lovell on the Files Covering Contact with Soviet Scientists and Visits to the Soviet Union, 1963 File.
[1.90] Letter from Frederick H. Murphy to I.S. Shklovsky on 9 January 1963, Library of Congress, Carl Sagan and Ann Druyan Archive.

[1.91] Letter from Shklovsky to Sagan on 26 September 1963, Library of Congress, Carl Sagan and Ann Druyan Archive.
[1.92] Gilman, B. (1989) A History of the Deliberate Interference with the Flow of Mail: The Cases of the Soviet Union and the People's Republic of China : A Report. U.S. Government Printing Office, Washington, D.C.: vii.
[1.93] Gilman, B. (1989) A History of the Deliberate Interference with the Flow of Mail: The Cases of the Soviet Union and the People's Republic of China : A Report. U.S. Government Printing Office, Washington, D.C.: 1.
[1.94] Gilman, B. (1989) A History of the Deliberate Interference with the Flow of Mail: The Cases of the Soviet Union and the People's Republic of China : A Report. U.S. Government Printing Office, Washington, D.C.: 2.
[1.95] Pessen, E. (1990) Appraising American Cold War Policy by its Means of Implementation. Reviews in American History 18(4): 453.
[1.96] Pessen, E. (1990) Appraising American Cold War Policy by its Means of Implementation. Reviews in American History 18(4): 460.
[1.97] United States. Congress. House. Committee on Post Office and Civil Service. Subcommittee on Postal Operations and Services (1988) Soviet disruption of mail: hearing before the Subcommittee on Postal Operations and Services of the Committee on Post Office and Civil Service, House of Representatives, One Hundredth Congress, second session. U.S. Government Printing Office, Washington, D.C.
[1.98] Gilman, B. (1989) A History of the Deliberate Interference with the Flow of Mail: The Cases of the Soviet Union and the People's Republic of China: A Report. U.S. Government Printing Office, Washington, D.C.: 4.
[1.99] Tabarovsky, I. (2019) Understanding the Real Origin of that New York Times Cartoon. Tablet Magazine. https://www.tabletmag.com/jewish-arts-and-culture/285781/soviet-anti-semitic-cartoons.
[1.100] Sagan, C. (1986) Contact. Simon and Schuster, New York.
[1.101] Shklovsky, I.S. (1991) Five Billion Vodka Bottles to the Moon. WW Norton & Co. Inc., New York: 251.
[1.102] Sagan, C.E. and Shklovsky, I.S. (1966) Intelligent Life in the Universe. Holden-Day, San Francisco: vii.
[1.103] Sagan, C.E. and Shklovsky, I.S. (1966) Intelligent Life in the Universe. Holden-Day, San Francisco: vii.
[1.104] Shklovsky, I.S. (1991) Five Billion Vodka Bottles to the Moon. WW Norton & Co. Inc., New York: 252.
[1.105] Shklovsky, I.S. (1991) Five Billion Vodka Bottles to the Moon. WW Norton & Co. Inc., New York: 252.
[1.106] Sagan, C.E. and Shklovsky, I.S. (1966) Intelligent Life in the Universe. Holden-Day, San Francisco: vii.
[1.107] Sagan, C.E. and Shklovsky, I.S. (1966) Intelligent Life in the Universe. Holden-Day, San Francisco: viii.
[1.108] Shklovsky, I.S. (1991) Five Billion Vodka Bottles to the Moon. WW Norton & Co. Inc., New York: 252.

[1.109] Burr, W. (2006) How Many and Where Were the Nukes?: What the U.S. Government No Longer Wants You to Know about Nuclear Weapons During the Cold War. National Security Archive Electronic Briefing Book No. 197. https://nsarchive2.gwu.edu/NSAEBB/NSAEBB197/index.htm.
[1.110] By some estimates, more than 100 nuclear detonations in even a best-case-scenario for attack by an aggressor state would result in devastating environmental impacts, sometimes called a "nuclear winter" or "nuclear autumn". For more, see: Pearce, J.M. and Denkenberger D.C. (2018) A National Pragmatic Safety Limit for Nuclear Weapon Quantities. Safety 8 (25).
[1.111] Sagan, C.E. and Shklovsky, I.S. (1966) Intelligent Life in the Universe. Holden-Day, San Francisco: 358.
[1.112] Sagan, C. (1975) Communication with Extraterrestrial Intelligence. MIT Press, Cambridge.
[1.113] Sagan, C. (1983) Nuclear War and Climatic Catastrophe: Some Policy Implications. Foreign Affairs 62(2): 292.
[1.114] Shuch, H.P. (2005) Prof. Phillip Morrison. The SETI League. http://www.setileague.org/admin/philmorr.htm.

2

Reading the Cosmos: What Our Science Tells Us About the *Science* of Another World

Carl L. DeVito

Department of Mathematics, University of Arizona, Tucson, Arizona, U.S.A.

Abstract
Some years ago, the author and a colleague constructed, as a thought experiment, a language based on the fundamental facts of science. It has been suggested that this language might be suitable for interstellar communication. We briefly discuss this language and then examine how the science of another world might differ from our own. Our strong reliance on the sense of sight and our evolutionary history, as both predator and prey, have influenced our understanding of space and time. This understanding plays a fundamental role in Newtonian mechanics and in both differential and integral calculus. Investigating this role enables us to identify aspects of our science, and of our mathematics, where we and an alien race might differ.

Keywords: Interstellar communication, newtonian mechanics, calculus, Euclidean geometry, hyperbolic geometry

2.1 Introduction

We don't know if alien societies exist and, if they do, what they might be like. Our method of searching, however, does put restrictions on the kind of society we might find. We will only detect societies that have a technology. It seems likely that such a society will also have a science on which to base that technology. Science here on Earth has shown us certain obstacles that had to be overcome before progress could be made. An alien society,

Email: devito@math.arizona.edu

Anamaria Berea (ed.) Technosignatures for Detecting Intelligent Life in Our Universe: A Research Companion, (33–46) © 2022 Scrivener Publishing LLC

in developing its science, will have faced the same obstacles. It would seem then, that by examining the development of science here, we can gain insight into the possible nature of the science on other worlds.

2.2 The Why and How of Interstellar Communication

The stars, as we all know, are very far away. Reaching them by spacecraft is well beyond our current abilities and will remain so for the foreseeable future. So, if we are to contact an extraterrestrial society, we must do it by radio or some other type of electromagnetic radiation.

This, of course, limits the kind of civilization we can hope to find. There may be many races out there that excel in art, music, philosophy or literature that cannot, or do not choose to, send radio waves into space. At present there is no way we can detect such a race.

There is so much we'd like to know about any race we do contact. What do they look like? Do they have two sexes? What are their social organizations like? Do they have music, art, theological ideas, political ideas, scientific theories?

Unfortunately, there is no easy way to obtain such knowledge.

I think the desire for such knowledge is so great that many forget the incredible technical challenges this kind of communication involves. We may be able to do little more than recognize each other's existence. But, if nothing else, interstellar signaling is at least a way to pass on a "legacy" to other intelligences in the galaxy.

It is a way to let them know we were/are here, that we lived and learned and did things, and experienced life. It is one way of extending the life span of humanity and a necessary one because the Earth will not last long by cosmic standards and once it is gone, so is all human accomplishment. Presumably, the society we contact will do the same, giving us some knowledge of their totally alien civilization. We want to know about them just as we want to know about the Mayans, Egyptians, Greeks, and Romans. Not because this knowledge will "save" or even change us, but because they lived. Just knowing they exist will enrich our thinking and broaden our understanding of the nature of life.

I do not see contact with an alien society as a "metaphysical" experience, nor do I expect that a society that knows nothing about us will somehow have the answer to all our problems.

Curiously, there is a persistent belief that communicating with an alien society will be easy—almost trivial! I have spoken about such communication at many scientific conferences. Invariably someone will insist that pictures, or music, or computers will easily handle the problem and have us and the aliens talking like old friends.

I find this rather puzzling. Wouldn't it mean that human to human communication would also be trivial? So why do people bother learning foreign languages? Couldn't we travel the world with a sketch pad (or a musical instrument if you can play one) and instantly communicate with the natives in any country? As for computers, why haven't they taught us to speak "dolphin"? In connection with language, even human languages, our computers still have their limits.

We must keep in mind that when two humans communicate, they both have human bodies with similar needs, they share the same sense organs and also share the culture of their time. Even with all this, face-to-face communication can be difficult, as anyone who has done any travelling has learned.

Now imagine communicating with a non-human. We recall a statement usually attributed to Wittgenstein: "Even if a lion could speak our language, we wouldn't understand him." This is not saying that a human and a lion cannot communicate. I mean, if a lion came into your room and said "I just woke up from a ten-hour nap and I am famished. I could really go for a snack right now," I think you would get the message. What Wittgenstein's remark is saying is that even with a common language two beings can be so different that only a very low level of communication is possible. This may be the case between us and an extraterrestrial race. We must be aware that the level of communication possible with such a race may be frustratingly low. After all, we may have more in common with a lion than we have with an alien.

Interstellar communication will not be a dialogue in the usual sense. If we hear from a society that is, say, ten light-years away, we will not send a message then wait twenty years or more for a reply, then send another message, and so on. What we should, in my opinion, do, and what I believe the aliens will also do, is immediately send a series of messages, perhaps a week or a few days apart, each containing more information than the last, but each containing some redundancy as a check on understanding. In this way each society leaves with the other a knowledge of its existence and, perhaps, something of its accomplishments.

My allusion to the ancient peoples listed above shows why we need social scientists on any "SETI team." This kind of communication will be, at least for quite a while, one-way, as indicated above. Our communication with past societies on Earth is one-way. We will have to glean what we can from the messages the aliens send us, extracting some understanding of their sense organs, their psychology, and their social structures along with whatever we learn of their physics, chemistry and mathematics.

But what should our messages say? More to the point, what can we say that would be comprehensible to an extraterrestrial?

2.3 An Early Language

If the aliens are mentally human-like we may be able to communicate by creating a language based on logic. The idea of constructing such a language goes back at least as far as Gottfried Leibnitz (1646-1716). Late in the nineteenth century Giuseppe Peano (1858-1932) discussed such languages extensively. These efforts had an important impact on the development of mathematical logic, and computer scientists, in connection with the development of programming languages, also drew on this work. However, the idea of using a logic-based language as a means of communication, the original intent of both Leibnitz and Peano, was forgotten until 1960. In that year the Dutch mathematician Hans Freudenthal (1905-1990) published a remarkable book called *Lincos* (a contraction of the words Lingua Cosmica). He set himself the problem of designing a language suitable for communicating via radio with (hypothetical) intelligent extraterrestrials. Since such beings would have no knowledge of our natural languages and could not be shown physical objects or demonstrations, there is nothing but logic on which to base a common language. Freudenthal showed that, if their thought processes are human-like, a language suitable for mutual communication could be taught to them. The disadvantage here is in the length of this work. A good deal of the book is devoted to developing the language. He even planned a second volume but this was never completed. References to the works quoted above can be found in Freudenthal's book *Lincos – Design of a Language for Cosmic Intercourse*, which is available online.

2.4 A Language Based on Science

My colleague Richard T. Oehrle, a linguist and logician, and I tried a different approach to cosmic discourse [2.1]. The society we detect will have something like a radio telescope, hence they have an interest in their astronomical environment.

Our astronomers know a lot about stars of course, but they have very detailed knowledge of only one star, the sun. They would love to get the same detailed knowledge about some other star. Our geologists and meteorologists would love to get detailed knowledge of some other planet, and our space scientists would love to get detailed knowledge of some other "solar system". An alien society with an interest in its astronomical environment would welcome such knowledge from us. In other words, the first step in this communication is to develop the ability to exchange *precise, quantitative scientific information* about our respective planets, stars and solar systems, because these are things of mutual interest.

We may never get beyond this, admittedly low, level because, as we indicated above, it just might be that the races are too different. But this level is far better than most messages that have been sent. Many of them amount to: "Hello, here's a picture. I hope you can figure out what it means".

But let me make a strange statement that might clarify a serious, underlying problem that arises here. Suppose our aliens said that their planet had a mass of so many zwangs, is located at a distance of so many waylets from their star, which emits so many quibbets of energy per soquit.

This statement tells us nothing because we don't know what the units mentioned mean. The aliens would have a similar problem if we spoke of grams, kilometers, and calories. So, the first task, if this kind of exchange is to be achieved, is to figure out ways to communicate the human units of measurement which, of course, are just conventions. This is what Oehrle and I did in our paper. We used universal facts of science to try to communicate our units.

Our language begins with simple arithmetic and uses the numbers, 1. 2. 3. ... to introduce some simple linguistic concepts. It is based on the assumption that a race with the radio telescope knows how to count. This gives us the ability to present the periodic table of elements. These elements are universal (at least in the visible universe) and the periodic variation of their properties is an observable fact. This enriches the language and greatly expands its expressive power.

We discussed our system of atomic weights and stated Avogadro's law. This law is crucial to any advance in chemistry and using it we can communicate our gram. Chemical elements react and these reactions involve heat exchanges which enables us, since we have defined the gram, to define our calorie. The distinctive way metals melt can be used to define our temperature scale. We used the Kelvin scale in our paper since we want to discuss the gas laws and introduce the concepts of pressure and volume. A science of chemistry requires a knowledge of all these things and we are assuming that the aliens have at least a rudimentary science of chemistry. We cannot, of course, guarantee that this will be understood, but it might be, and the "payoff" would be a great advance in our scientific knowledge.

I want to emphasize that Oehrle and I did not simply say that the things sketched above could be done. We did them in detail, crossing all the t's and dotting all the i's. These details can be found in our paper. There we also made our assumptions very clear but, let me stress, we made no assumptions about the structure of the alien society nor any about the sense organs of their members.

It is widely believed that the sense of sight will be universal and efforts have been made to exploit this by introducing pictures into our communication. This must be done with great care. Pictures may easily be misinterpreted [2.2]. Oehrle and I briefly discussed at the very end of our paper how we might use pictures.

2.5 Our Senses, Our Perceptions and Our Science

My purpose in this section is to try to get some idea of how an alien science might differ from our own. The differences may be quite striking. I think our evolutionary history and the peculiarities of our senses have had a profound effect on the development of our science here on Earth. An alien race may have a science. Any such race that we *can detect* probably does have a science. But the beings we encounter will probably have a different evolutionary history and, even if they see and hear, they may see and hear in a way that is different from us. I will try to clarify these statements in the discussion below.

I hope my remarks will stimulate a discussion of these matters among my colleagues in both the physical and the humanistic sciences. This may be a way to get some idea of how an alien race may differ from us in very subtle ways.

2.5.1 Motion

Our intuitive understanding of motion may have come from our history of being both prey and predator. When jumping for a branch in order to escape a predator you cannot stop to think about it. You must do it automatically, and do it right. Similarly, when you throw your spear at a fleeing animal you must anticipate where to throw the spear so as to reach the point where the animal will be when it gets there. Again, you cannot think about it, you have to internalize the properties of motion. Many of us still have this intuition. One sees it in students who find Newtonian physics and differential calculus natural and appealing.

To give a precise description of motion we need the idea of velocity (speed plus direction) and that of acceleration. Bear in mind that average velocities are not what we need here. In a drive of 100 miles made, say, in two hours, our average speed is 50 mph. But that is not what our speedometer reads. The speedometer gives our velocity at each instant of the drive. One takes the average speed over smaller and smaller intervals of time, hence over smaller and smaller distances and obtains a limiting value. So, we need to have an understanding of both space and time "in the small."

2.5.2 Space "in the small"

To get an idea of how we perceive space "in the small" consider how we define the volume of a soup can. We start by defining the volume of a cube. Then we fill the can with cubes, say the size of dice, and add the volumes of these. This gives only a rough approximation to the volume because the flat sides of the die don't fit too well in the round can. So, we fill it with smaller cubes, say each one-tenth the size of a die. This gives a better approximation because the smaller cubes fit better. We can easily imagine this process continuing to smaller and smaller cubes, cubes each the size of a grain of sand or even smaller. The numbers one gets do not fly around aimlessly. They converge to a fixed number we take as the volume of the can. But note how here we think of space as infinitely devisable. The ideas sketched above can be made precise. They lead to integral calculus.

2.5.3 Time

Our perception of time changed radically over the centuries [2.3]. The Roman philosopher Martianus Capella (360-420) suggested that there might be atoms of time, each atom having a small, but positive, duration. This idea seems to have been quite popular. We know that it was discussed by the Islamic scholars and the word for "instant" in the Italian language is "attimo"; pronounced like their word for atom. The first person to assign a duration to an atom of time, however, was the famous Rabbi Maimonides (1138-1204). In his book *A Guide for the Perplexed* written in 1190 he suggested that there are $(60)^{10}$ such atoms in an hour; 60 minutes in an hour, 60 seconds in a minute, perhaps 60 subseconds in a second, etc. Why stop at ten? As far as I know, the rabbi never gave his reasons.

In this model each atom of time had a unique predecessor and a unique successor. This is not true of the current model. Here time is a continuum. It consists of duration-less instants. Any interval of time can be subdivided into as many smaller parts as one wants.

I should mention that modern physicists have discussed the idea of an atomic time. They call this hypothetical atom a "chronon".

By the time of Galileo (1564-1642), and perhaps much earlier, the modern idea of continuous time seems to have been widely accepted.

Why did the old idea fade from popularity especially so long ago when there was little mass communication? It may have been the growth of music and the universal popularity of this art form that led to the change. Our sense of hearing then combined with what we find audibly pleasing may have been responsible for our current understanding of a continuous time. Without this view our investigation into the properties of motion might have been seriously hampered.

Combining our perceptions of a continuous space and a continuous time enabled us to formulate the idea of instantaneous velocity and instantaneous acceleration. In the terminology of mathematics what we needed was the concept of the derivative. This concept is central to a precise discussion of motion, and hence central to both physics and engineering.

The laws of motion are expressed as equations involving derivatives, and in our formulation of them we come upon another important aspect of human perception.

2.5.4 Space "in the large"

Many ancient peoples had an understanding of space. The annual flooding of the Nile obliterated the boundaries between adjacent farms and the Egyptians had to figure out ways to correctly reset them after the water receded. Archaeological research in the twentieth century shows that in Mesopotamia, as early as 1800 B.C., the Pythagorean relation between the sides of a right triangle was known [2.4]. But the first systematic study of space was carried out in ancient Greece and has come down to us in a series of books written about 300 B.C. by Euclid. This is the geometry most of us studied in high school. Many of the results stated, at least in the beginning, seem intuitively obvious to most people. Immanuel Kant thought that this geometry was hardwired into our brains and was the essence of how we see [2.5; p. 63].

In the nineteenth century a new geometry was introduced. This has come to be called hyperbolic geometry. It was published independently and more or less simultaneously by the Hungarian Janos Bolyai (1802-1860) and the Russian Nicolai Ivanovitch Lobachevsky (1793-1856). This stirred a great controversy in the mathematical community. How dare these young men question Euclid! Anyone could see that Euclidean geometry was the obviously the true description of space. Many were sure that this new geometry could not be consistent.

The controversy died when the Italian mathematician Eugenio Beltrami (1835-1906) showed that the consistency of Euclidean geometry, which no one doubted, implied the consistency of the new one [2.5; pp. 61-71]. They are actually equiconsistent. Assuming the consistency of either one enables you to prove the consistency of the other.

But now the physicists had a problem. They had assumed, along with everyone else, that Euclidean geometry was the "true" description of the space of our experience. But now they knew that it was just one possible "model" for space. Which model was "right"?

In response to this question the great French mathematician Henri Poincare (1854-1912) had this to say:

> All measurement involves both physical and geometrical assumptions, and the two things, space and matter, are not given separately, but analyzed out of a common experience. Subject to the general condition that space is to be changeless and matter to move about in space, we can explain the same observed results in many different ways by making compensatory changes in the qualities we assign to space and the qualities we assign to matter. Hence, it seems theoretically impossible to decide by any experiment what are the qualities of one in distinction from the other [2.6].

So, we must choose a model for space, a geometry, and then formulate the laws of motion in that model. Many geometries are now known but the pioneers in the study of motion, Galileo, Kepler, and Newton, knew only the geometry of Euclid. This was fortunate for two reasons.

First, the associated trigonometry is especially easy. The extremely useful trigonometric functions, sine and cosine, were studied and tabulated very early [2.7]. Using these functions astronomers could compute the distance of a planet from the sun divided by the distance of the Earth from the sun. By taking the latter distance as their unit, they got an idea of the relative size of the solar system [2.10, pp. 57-60]. Secondly, the equations of motion are simplest when they are formulated in the geometry of Euclid [2.7]. It seems we are very lucky to see things as we do.

To get a better idea of what I mean by seeing in a Euclidean way, let me give an example. Recall that two lines are parallel if they lie in a plane and never meet.

Given a line and a point not on that line, if you ask a class of college students, or a class of senior citizens, "Can we draw a line through the given point that is parallel to the given line?" they immediately answer: "Yes".

If you then ask: "How many such lines can you draw?" there is usually a moment's hesitation, but then an answer: "One, just one."

This simple statement that many find "obvious" has a striking consequence. From it one can prove the remarkable fact that in the geometry of Euclid the three angles in *any* triangle always add up to 180 degrees. The size of the triangle doesn't matter.

This is not true of the other geometries and is one reason why the trigonometry is so simple in Euclid's geometry. As already mentioned, the laws of motion are especially simple when formulated in this geometry [2.8].

The fact that the Euclidean model conforms to how we see means that as we studied the solar system and formulated the laws of motion the mathematics was simple. Had we "seen" differently we could still have made progress but it would have been hampered by a more difficult mathematics. An alien race that does not see "Euclidean" may have a very different formulation of the laws of physics [2.7].

Let me stress that since there are other geometries that are equiconsistent with that of Euclid, there is no reason why another race may not choose one of them as their model for space, as emphasized by Poincare in his remark quoted above.

Many animals on Earth have the same senses as we do but they seem to rely more on the sense of smell as opposed to sight. Dogs and cats come to mind. Imagine a race of dogs that somehow develops intelligence equal to that of humans. If they have a science, I think it would begin with chemistry. The early part of this paper might be understandable to such a race. It is not clear what the long-range sight of a dog is, but it is conceivable that it is non-Euclidean. If so the trigonometry and physics of motion of this race of intelligent dogs, if they had these things, would look very different from ours. The difficulty of their mathematics may hamper their progress in these areas.

2.5.5 Mathematics

Many believe we would share a great deal of mathematics with any race that has a technology. Let us take a closer look at this belief.

There is no denying that mathematics is useful in the physical sciences. It also has a disconcerting habit of becoming useful, sometimes long after it was discovered (or created; see below). A striking example is the system of complex numbers introduced by Geralomo Cardano (1501-1576) and Rafael Bombelli (1526-1572) during the Italian Renaissance. These numbers now play a fundamental role in quantum physics and in certain areas of electrical engineering; two subjects that didn't exist until several centuries after the Renaissance.

But why is mathematics so useful? Is it just a reflection of human ingenuity or is something more involved?

The German physicist Heinrich Hertz (1857-1894) was the first to produce electromagnetic waves in the laboratory, thereby confirming Maxwell's prediction, made twenty years earlier, that such waves should exist. The first indication that such waves existed was the fact that from Maxwell's equations, governing electromagnetic phenomena, one could deduce the wave equation.

Regarding these equations Hertz remarked:

> One cannot escape the feeling that these mathematical formulas have an independent existence and an intelligence of their own, that they are wiser that we are, wiser even than their discoverers, that we get more out of them than was originally put into them. [2.5; p. 98].

He seems to be saying that, regarding our second question, mathematics has an existence that is independent of its human creators; he refers to them as "discoverers". This is the view of the Platonists school (see below).

In regard to our first question Eugene Wigner (1902-1995), winner of the Nobel Prize for Physics in 1963, had this to say:

> The first point is that the enormous usefulness of mathematics in the natural sciences is something bordering on the mysterious and that there is no rational explanation for it. [2.9].

There are other views of mathematics that are worth noting here.

Many see the subject as a kind of "cookbook," with recipes for solving problems of various kinds, and rules for carrying out calculations. These things are important but they are only the most obvious aspects of the subject; many of them can now be done by our computers. But most of mathematics involves abstract reasoning; thinking logically and critically in abstract terms.

The world of mathematics is vast and constantly growing. It is not the world of our everyday experience. It is a world of ideal objects. A mathematical point is zero dimensional. Saying it is a tiny dot with extremely small, but positive, radius leads to a contradiction. Nothing in the real world is one dimensional, but a mathematical line is. And the thickness of a mathematical plane, unlike that of a sheet of paper, is zero.

But how did the human race come upon this world?

Some say mathematical objects exist in some kind of meta-world. When a mathematician defines a mathematical object, he or she is precisely describing an object that already exists somewhere "out there". Those with this view are called "extreme Platonists." So, in this view, mathematicians *discover* mathematics.

If the Platonists are right, we would expect an alien race to also discover some mathematics since it is so useful in science. Furthermore, the mathematics of that race could very well be the same as that of the human race.

There is another school of thought here. The "strict formalists" insist that mathematical objects exist only in the human mind. If the formalists are right, an alien race, even one that has a science, might not have a mathematics as we understand it. Our mathematics would be unknown to them and, since our physical theories are based on this mathematics, they, too, would be unknown to this race.

We again quote Wigner: "Second, it is just this uncanny usefulness of mathematical concepts that raises the question of the uniqueness of our physical theories [2.9]."

2.6 Conclusion

How much of our science is universal? Would an alien science mirror our own? In the first part of this paper, we emphasized the universality of chemistry. The elements are the same throughout the visible universe, so alien chemists would be studying the same elements we study here. Whatever the manipulative organs of an alien might be they cannot handle individual molecules. A system of atomic weights is needed to make progress in chemistry and linking these weights to the weights of large quantities of matter is crucial if one is to make meaningful calculations. So some form of the Avogadro law must be known.

Our basic physics—I mean Newtonian mechanics—might be a little more questionable. Not the laws of motion themselves, but the way they are formulated. If our concepts of continuous space and continuous time are shared, aliens may formulate velocity and acceleration as we do. How they formulate the equations of motion will depend, as indicated above, on their geometry. As for our more sophisticated physical theories, say relativity or quantum mechanics, we refer to Wigner's statement quoted above. Any extraterrestrial race we contact will probably have a science, but just how much their science and our science will have in common is not entirely clear.

References

[2.1] DeVito, C. L. and R. Oehrle (1990) A Language Based on the Fundamental Facts of Science. *Journal of the British Interplanetary Society* 43: 561-568.

[2.2] Vakoch, D. A. (2000) The Conventionality of Pictorial Representation in Interstellar Messages. *Acta Astronautica* 46, no. 10–12: 733-736.

[2.3] Whitrow, G.J. (1961) The Natural Philosophy of Time. *Thomas Nelson & Sons Limited, London.*

[2.4] Edwards, H. M. (1984) Galois Theory. *Springer-Verlag, New York.*

[2.5] Osserman, R. (1995) Poetry of the Universe—A Mathematical Exploration of the Cosmos. *Anchor Books a division of Random House, Inc., New York*

[2.6] Poincare, H. (1952) Science and Hypothesis. *Dover, New York.*

[2.7] Lamphere, R.L. (2002) Solution of the Direct Problem of Uniform Circular Motion in Non-Euclidean Geometry. *The American Mathematical Monthly* 109: 650-55.

[2.8] McLeary, John (2002) Trigonometries. *The American Mathematical Monthly* 109: 623-38.

[2.9] Wigner, E. (1960) The Unreasonable Effectiveness of Mathematics in the Natural Sciences. *Communications in Pure and Applied Mathematics,* vol. 13 no. 1. *John Wiley & Sons Inc., New York.*

[2.10] Kaufmann, William J. (1994) Universe 4th Edition. *W.H. Freeman and Company, New York.*

3

The Impact of Discovering the First Technosignature

Ross Davis[1*], Kevin Schillo[2] and Lucianne Walkowicz[3]

[1]*Indiana University-Purdue University Indianapolis (IUPUI), Indianapolis, Indiana, US*
[2]*University of Alabama in Huntsville, Huntsville, Alabama, US*
[3]*The Adler Planetarium, Chicago, Illinois, US*

Abstract
One important question in the search for extraterrestrial intelligence (SETI) is what would the impact be to us here on Earth if we discovered that we are not alone in the universe? Even though extraterrestrial life has not been discovered yet, the potential answer to this question has been considerably speculated about due to persistent scientific and popular interest in SETI. A main goal here is to be able to better answer this question, especially if the discovery involved the first-ever technosignature, which is a sign of extraterrestrial technology such as a radio or optical signal broadcast through space. Doing so can better inform us (especially those handling the scientific confirmation and public disclosure of such a discovery) of the potential benefits and risks associated with the societal impact of the discovery. And even if a discovery has not been made yet, it can indirectly inform as to the impact of our own destiny into space. To this end, a historical and sociological approach is used, where past historical events that alleged the discovery of extraterrestrial life (events such as inconclusive scientific discoveries and science fiction stories) are used as an analogy. In conjunction with this, a model of the societal impact of discovering the first confirmed technosignature is presented, based on a process representing the discovery's hypothetical diffusion into society. This is so as to understand what the extent of the impact can be of an actual discovery.

Keywords: SETI, technosignature, discovery, disclosure, exoplanets, social impact, UFO

**Corresponding author*: rdavis70@protonmail.com

Anamaria Berea (ed.) Technosignatures for Detecting Intelligent Life in Our Universe: A Research Companion, (47–56)

3.1 Introduction

While the decades-long SETI has not resulted in the discovery of alien life yet, which can lead to skepticism or impatience as to whether alien life would ever be discovered at all, scientific and technological discoveries still occur when we least expect them, such as the serendipitous discovery of cosmic background radiation lending support for the Big Bang theory [3.10]. And considering that thousands of confirmed exoplanets have been discovered in our galaxy, with many more expected along the way, there is much ground still to cover as far as SETI is concerned. Transcending this is the fact that interest in discovering alien life has been of significant societal interest across modern history, as highlighted in the next section through analogous historical events. The section after this one proposes a qualitative model of the potential societal impact of the first technosignature discovery (or other type of alien discovery), based on the discovery's prospective diffusion into society. While the analogous historical events provide case study or anecdotal insights into the potential impact, the model serves as a systematic conceptual framework that can be applied across time and space, whether to past events or future scenarios.

In sum, this can better inform us (especially those handling the scientific confirmation and public disclosure of such a discovery) of the potential benefits and risks associated with the societal impact of the discovery. And even if a discovery has not been made yet, it can indirectly inform as to the impact of our own destiny into space.

3.2 Cultural Impact Based on Analogous Historical Events

Many believe that proof of extraterrestrial intelligence would constitute the greatest discovery in history. How society reacts to such a discovery will depend on the nature of the discovery, the technological advancement of the alien society, and whether the extraterrestrials are benevolent or malevolent toward humanity. For example, a radio transmission originating from a civilization located more than one thousand light-years from Earth would undoubtedly have a far less significant cultural impact than if extraterrestrial spacecraft were to land in major cities all over the world. To study the potential cultural impact of extraterrestrial contact, it is useful to examine public reaction to hoaxed or unconfirmed discoveries of alien life.

An infamous example was a radio adaption of the H.G. Wells novel *The War of the Worlds.* As one of the earliest novels featuring alien contact, it depicted an invasion of New Jersey by hostile extraterrestrials from Mars. The infamous radio adaptation was narrated by Orson Welles and broadcast in the United States on October 30, 1938. Contrary to popular belief, there was not mass hysteria during the broadcast. This was partly due to the broadcast's comparatively small audience, with only a small subset of the audience believing that the radio drama was real [3.14].

Another noteworthy event was the examination of Allan Hills 84001, a meteorite originating from Mars discovered in Antarctica in 1984. In 1996, McKay *et al.* published a paper in *Science* that presented possible signs of ancient Martian microbial life in the meteorite [3.7]. This attracted a large amount of media attention and even prompted then-President Bill Clinton to hold a press conference to discuss the meteorite [3.1]. Film footage of Clinton's press conference about the meteorite was even used in the movie *Contact*, making it appear as though the president was commenting on the alien message in the film. Ultimately, the scientific community determined that the structures observed in the meteorite were not unambiguous proof of microbial Martian life [3.7].

3.3 Modeling the Impact Based on the Discovery's Diffusion into Society

The extent of the societal impact of the first technosignature discovery (if one was ever discovered and publicized) can be understood based on the diffusion of innovations theorized by sociologist and communication theorist Everett Rogers. Rogers developed the diffusion of innovations in connection with assessing the impact of development programs in agriculture, family planning, and public health across the world during the early part of the twentieth century. According to Rogers, the diffusion of an innovation into a society is a process starting with the discovery's adoption by its discoverers, then (through a ripple effect) by early adopters, followed by the early majority, late majority, and laggards [3.11]. A model of this diffusion of innovations applicable to the discovery of the first technosignature, so as to understand the discovery's impact, is as follows:

1. Discoverers - The SETI researchers (or researcher) who claim to have discovered the first-ever technosignature.
2. Early adopters - The discoverers' peers in the SETI research community who subsequently adopt the claim, especially after scientifically confirming it through a peer review process.
3. Early majority - Members of the broader scientific community beyond SETI, government officials (especially those with policy-making, public relations roles), and opinion leaders (such as science communicators like the late Carl Sagan) who adopt the claim after the early adopters.
4. Late majority - Those among the mass media, other societal institutions, and the general public.
5. Laggards - People who have traditional belief systems, or lack interest in the subject.

The aforementioned process of diffusion, being in context of a model, is a general representation that would not depict all possible scenarios. For example, it is possible (though not considered appropriate) for the discoverers to promote their claim by largely bypassing the scientific peer review process and going straight to the mass media, thereby confounding who the early adopters would be regardless of whether the claim is scientifically valid. The pace and scope of diffusion of the first technosignature discovery as modeled as such would depend on the technosignature itself, communication channels, the social system, and time [3.11].

As far as the degree of diffusion depending on the technosignature itself, one tool to gauge this is what is known as the Rio Scale. The Rio Scale is designed to communicate the significance of a SETI signal to the general public based on a score ranging between 0 and 10, with 10 representing an extraordinary discovery (such as a technosignature intended for Earth and originating within our solar system). The score is determined by a formula that accounts for the potential consequences and credibility of a signal being truly extraterrestrial. An online calculator has been developed that SETI researchers (and the general public) can access to score a SETI signal relative to the scale; the calculator employs a questionnaire to input responses pertaining to observational characteristics of the SETI signal. The scale has evolved over time, progressing through three versions, from Rio 1.0, to 1.1, to 2.0. Also, the scale is called the Rio Scale based on its having been presented at the 51st International Astronautical Congress in Rio de Janeiro in 2000 [3.3].

Regarding diffusion depending on communication channels and the social system, the communication channels would encompass in-person, print, and electronic form. In-person communication can be advantageous in conveying information in a fluid and personable manner to small groups, such as a group seminar. Print communication can be advantageous in conveying information in a fixed and reusable manner to a group of any size, whether a small research team or an international audience; this can include a print journal article. Electronic communication can, depending on the circumstances, incorporate advantages of both in-person and print communication, such as an audiovisual webinar that includes text-based chat and a recording option. Electronic communication, especially in the Information Age using the internet, would likely be the prominent communication channel, especially in the first technosignature discovery diffusing across the early and late majority, and on to laggards (based on the aforementioned diffusion model).

The social system manifests structure and functions that mediate the diffusion of the discovery, especially in terms of the institutions that comprise the social system. The initial institution would be the SETI research community, involving the discoverers of the first technosignature processing the discovery, followed by the broader scientific community beyond SETI, the government, the mass media, and so forth. These institutions would exert influence on the diffusion to an extent. These influences include (but are not necessarily limited to) policies, procedures, analyses, and resources that mediate the discovery's adoption. Examples of policy and procedures are the SETI protocols set forth by the International Academy of Astronautics (IAA) SETI Permanent Committee [3.5]. If there is an institutional environment whereby these influences are favorable overall toward the adoption of the discovery (if the discovery itself has scientific merit), then the pace and scope of diffusion of the discovery into society would likely be significant, and likewise the discovery's ultimate impact. On the other hand, an often-cited obstacle to this would be a SETI approach that risks being anthropocentric in understanding ET intelligence, for example: a technosignature containing mathematical information based on Euclidean geometry instead of the concept of infinity and prime numbers associated with our terrestrial mathematics [3.8], or originating from an exoplanet where the solvent of life would be a sulfur compound instead of water [3.13].

The diffusion of the discovery would also depend on how much time has passed since the discovery. Drawing upon the approach of science philosopher Thomas Kuhn, the discovery of the first technosignature would determine whether today's "normal science" would be altered tomorrow by a revolutionary one, based on the realization that we are no longer alone in the universe, especially in terms of intelligent life. Such a revolutionary science can be on par with the Copernican revolution, which was a paradigm shift from the Earth being deemed the center of the universe (the Ptolemaic model) to the sun being the center of our solar system (the heliocentric model) [3.4][3.6]. Historically, such revolutions become established gradually over time, for instance: the Copernican revolution began in the sixteenth century and was subsequently advanced through contributions from Galileo's observational astronomy in the seventeenth century, and Newtonian mechanics in the eighteenth century [3.6][3.15]. And even if the first technosignature would not be discovered yet, SETI can still yield discoveries in the interim that help to advance science and technological innovation (such as advances in astrobiology, astronomy, and other relevant academic disciplines; also, in the instruments and methods related thereto).

3.4 Multimodal Diffusion into Knowledge Systems of Complex Publics

Today, the detection of any technosignature faces a considerably more complex information landscape than the one conceived of by Rogers: for example, "electronic communication" now encompasses an extremely wide variety of modes, including person-to-person communications between familiars, social media (which facilitates information transfer not only between individuals, but institutions and organizations, and even non-sentient agents such as algorithms). The model of diffusion set forth in Section 3.3 represents a linear process, where insiders (i.e., the discoverers of the first technosignature) have perfect control of their information, allowing it to diffuse out to ever-less-specialist publics beyond the initial discovery team.

However, rarely does even the tamest scientific discovery traverse such a linear path; consider, for example, the information environment of a typical American Astronomical Society (AAS) conference. The AAS conferences take place twice a year, in winter and summer, and are amongst the largest annual gatherings of professional astronomers in the United States. Researchers prepare talks and posters to announce new scientific results at the AAS conference as part of its scientific program, and accompanying press conferences are often also held to announce results that are particularly high profile (where significance is judged based on myriad factors, including not only the scientific import of the result, but also the relevance of the instrumentation used to national agencies' agendas, and even the perceived level of interest from the broader publics who are not professional scientists). In general, new discoveries are announced not only in peer-to-peer communication venues within the conference program, but also in academic journal articles (which appear both behind the paywalls of subscription-based journals, and on publicly accessible preprint archives), in press conferences and the associated popular science press written following the press conference, and on social media outlets (where they are posted by scientists themselves, as well as by members of the press, media agencies, and public relations offices of both funding agencies and the institutions that are home to the scientists on the discovery team). In ideal practice, the use of press embargoes means that all of these modes are activated simultaneously; in reality, embargoes are broken regularly, both intentionally and accidentally, so word of new discoveries tends to leak through informal channels prior to being released "officially."

The multimodal nature of the modern information landscape disseminates information into an even more complex, diverse ecosystem of individuals' beliefs and existing knowledge. Denning *et al.* [3.2] provide an overview of the factors affecting how information of a potential technosignature might be received, and make particular note of the issues of trust (in scientists, in institutions, and between individuals) that further complicate how such a discovery might be received. For instance, while the Rio score might be helpful to those who know and understand it (e.g., currently the Rio Scale is most familiar to SETI scientists, and some members of the science press), it is not necessarily going to be seen as definitive by everyone. Because of the essential role that trust plays in the acceptance of new (and particularly controversial) information, one could argue that the Rio Scale would be seen as definitive predominantly by those who already have a high degree of trust in scientists.

Denning *et al.* [3.2] also helpfully elucidate the habits of thought in the SETI community that further influence the ways and means a technosignature discovery might be communicated beyond the initial discovery team. Thoroughly addressing how these habits of thought may affect the diffusion of a technosignature discovery could be its own volume, and thus is unfortunately beyond the scope of this chapter; however, we draw the reader's attention to the last item in their list, namely that there exists an:

> understandable tendency to think that humanity has always wondered if we are alone, and to reproduce that idea in service of SETI because it is a fascinating question to us— when in fact most of humanity in

> most of the world has never really thought they were alone, but rather lived in a cultural world with many other kinds of beings, lives, cultures, systems, and even worlds.

To make this statement specific to the questions of alien life that SETI hopes to address, a 2019 Gallup poll found that fully 75% of Americans believe that some kind of life exists on other planets, 49% believe that "people somewhat like ourselves" exist beyond Earth, and 33% of US adults believe that at least some UFO sightings are alien spacecraft visiting Earth. This latter statistic is remarkably robust across age, gender, and socioeconomic factors, with a slightly higher proportion (40%) of believers identifying as "no religion" rather than Christian (the survey did not provide information on belief across other faiths) [3.12]. While not addressed in this particular survey, it is also notable that in the history of congressional support for SETI funding, the place of alien life within religious belief played a significant role: Senator Jake Garn (R-Utah) advocated for SETI funding in part because intelligent alien life (including alien life visiting Earth) is accepted as fact in his Mormon faith [3.9].

In light of such a sizable fraction of the populace taking the existence of alien life (and technologically capable alien life, at that) as a given, one has to question whether framing the discovery of a technosignature as an "innovation" or "revolutionary" is wholly accurate, even if we count ourselves amongst those who would describe a technosignature discovery that way. Even for SETI scientists, who might be amongst those most excited to learn of the first technosignature, elation would quickly give way to a slew of other questions about the nature of the signal, its repeatability (or not), and the possibility that we might respond to it (or not)—all questions that SETI scientists already grapple with in a theoretical context.

While complex networks of information transfer (and actors therein) would almost certainly stymie any attempts to control how news of a technosignature discovery diffuses, those who work on SETI and hope to be in the position of trying to convey such news should not despair. The tangled fields of communication across cultures, of trust, belief, and legitimacy, all mediated by technology that is largely beyond the scope of our control, is nothing less than the local challenges of what SETI aspires to do at a cosmic scale. By considering these differences—and how they might play into a discovery being seen as extraordinary, or merely a confirmation of what one has been sure of all along—practitioners within SETI have an opportunity to embrace a more nuanced understanding of our own endeavors.

References

[3.1] Clinton, B. (1996) President Clinton statement regarding Mars meteorite discovery. https://www2.jpl.nasa.gov/snc/clinton.html

[3.2] Denning, K., Berea, A., Capovac, K., Forgand, D., Korpela, E., Oman-Reagan, M., Tarter, J., & Tennen, L. (2019) *SETI and Post-Detection: Towards a New Research Roadmap*. 70th International Astronautical Congress (IAC), Washington, D.C., October 21st. Paper ID: 52210. https://iafastro.directory/iac/paper/id/52210/summary/

[3.3] Forgan, D., Wright, J., Tarter, J., Korpela, E., Siemion, A., Almár, I., & Piotelat, E. (2019) Rio 2.0: Revising the Rio Scale for SETI detections. *International Journal of Astrobiology* 18(4), 336-344.

[3.4] Fry, I. (2015) The philosophy of astrobiology. In: *The Impact of Discovering Life Beyond Earth*. 1st edition. S. Dick, (ed.) Cambridge University Press, UK: 23-34.

[3.5] International Academy of Astronautics SETI Permanent Committee. (2010) Declaration of Principles Concerning the Conduct of the Search for Extraterrestrial Intelligence. https://iaaseti.org/en/protocols/

[3.6] Kuhn, T. (2012) *The Structure of Scientific Revolutions*. 4th edition. University of Chicago Press, Chicago.

[3.7] McKay, D., Everett, K., Gibson, E., Thomas-Keprta, K., Hojatollah, V., Romanek, C., & Zare, R. (1996) Search for past life on Mars: Possible relic biogenic activity in Martian meteorite ALH84001. *Science* 273(5277), 924-930.

[3.8] Narens. L. (1997) Surmising cognitive universals for extraterrestrial intelligence. In *Astronomical and Biochemical Origins and the Search for Life in the Universe*. C. Cosomvici, S. Bowyer, and D. Werthimer (eds.) Editrice Compositori, Bologna: 561-570.

[3.9] Nielson, K. (1971) People on Other Worlds. https://www.churchofjesuschrist.org/study/new-era/1971/04/people-on-other-worlds?lang=eng

[3.10] Orf. D. (2013) 10 Awesome Accidental Discoveries. https://www.popularmechanics.com/science/health/g1216/10-awesome-accidental-discoveries/

[3.11] Rogers, E. (2003) *Diffusion of Innovations*. 5th edition. Simon and Schuster, New York.

[3.12] Saad, L. (2019) Americans Skeptical of UFOs, but Say Government Knows More. https://news.gallup.com/poll/266441/americans-skeptical-ufos-say-government-knows.aspx

[3.13] Schulze-Makuch, D. (2015) The landscape of life. In: *The Impact of Discovering Life Beyond Earth*. 1st edition. S. Dick, (ed.) Cambridge University Press, UK: 81-91.

[3.14] Schwartz, A. (2015) *Broadcast Hysteria: Orson Welles's War of the Worlds and the Art of Fake News*. Hill and Wang, New York.
[3.15] Thoren, V. (1989) Tycho Brahe. In: *Planetary Astronomy from the Renaissance to the Rise of Astrophysics*. Part A: Tycho Brahe to Newton. R. Taton and C. Wilson, (eds.) Cambridge University Press, UK: 3-21.

4

Searching for Extraterrestrial Intelligence by Locating Potential ET Communication Networks in Space

Ross Davis

Indiana University-Purdue University Indianapolis (IUPUI), Indianapolis, Indiana, US

Abstract

There have been periodic efforts in recent decades to search for extraterrestrial intelligence (SETI), especially by trying to find an extraterrestrial (ET) radio signal or other technosignature in space. Yet, no such technosignatures have been found. Considering the vastness of space, finding such technosignatures has been described as trying to find a needle in a cosmic haystack. To help resolve this, two hypotheses are proposed to aid SETI researchers in narrowing the search for ET technosignatures, based on a network analysis approach to locate where in space potential ET communication networks would most likely be. A potential ET communication network can use exoplanets as communication access points (e.g., placing a communication satellite into a planet's orbit, and/or an antenna on a planet's surface). The approach uses a topology where exoplanets are represented as nodes, and the lines of average distance (generalized communication paths) between adjacent exoplanets are represented as edges; the nodes and edges form local and wide planetary networks. Using the approach and data visualization on exoplanet databases can highlight locations of potential ET communication networks in space; an example is provided based on a NenuFAR radio telescope SETI project. The first hypothesis posits that an ET technosignature would more likely appear in a potentially habitable solar system containing a high concentration of planets, wherein the planets function as communication access points to facilitate a potential ET communication network. The second hypothesis posits that an ET technosignature would more likely appear in a highly concentrated cluster of potentially habitable solar systems. Contributions to the SETI field can be

Email: rdavis70@protonmail.com

Anamaria Berea (ed.) Technosignatures for Detecting Intelligent Life in Our Universe: A Research Companion, (57–68) © 2022 Scrivener Publishing LLC

increased accuracy in finding ET technosignatures, increased accuracy in reaching a Schelling Point (a mutual realization of how we and an ET intelligence can find each other), and promoting interdisciplinary SETI research.

***Keywords*:** SETI, extraterrestrial intelligence, exoplanets, technosignatures, networks, information science, sociology, astrobiology

4.1 Introduction

Since around 1959 there have been periodic efforts on Earth to search for extraterrestrial intelligence (SETI) by way of detecting of extraterrestrial (ET) radio signals from outer space—a type of SETI known as communication SETI [4.21]. So far though, no such signals have been reliably found. Considering the vastness of space and the countless stars in it, trying to find an ET radio signal has been described as being similar to trying to find a needle in a cosmic haystack. Still, a number of researchers engaged in communication SETI remain undaunted as they pursue innovative ways in which to detect ET transmission signals coming from space, such as going beyond the radio spectrum to explore optical (think lasers) and infrared. More broadly conceived, ET radio, optical, and infrared space signals, as well as other technological, non-biological signs of ET intelligence, are collectively known as technosignatures [4.22].

To aid SETI researchers in narrowing the search for ET technosignatures, this paper proposes two hypotheses based on a network analysis approach to locate where in space potential ET communication networks would most likely be. These hypothetical networks would be capable of exhibiting ET technosignatures that include transmitted radio, optical, or other types of electromagnetic signals. In such an approach, a potential ET communication network has a physical topology wherein the planets that a communicative ET intelligence could use as communication access points (in its own solar system and beyond) represent nodes, and the lines of average distance (generalized communication paths) between such planets represent edges. A planet can serve as communication access point in various ways, such as by placing a communication satellite into orbit around the planet, or by stationing an antenna on the planet's surface.

The following scenario, or algorithm, illustrates how the hypotheses can be tested by applying data visualization in conjunction with network analysis of exoplanet databases (such as those developed from NASA's Kepler and TESS space missions); the hypotheses posited to be situated where ET communication networks would most likely be.

A. An astronomical chart, preferably 3-dimensional (3D), would be generated showing local planetary networks (LPNs) and wide planetary networks (WPNs). Each LPN consists of exoplanets within a potentially habitable solar system, where each exoplanet represents a node, and the line of the average distance between each exoplanet represents an edge. (An LPN would be a generalized communication path if there was any potential ET communication network.) Encompassing the LPNs would be the broader WPNs. Each WPN consists of clusters of potentially habitable solar systems, where each of these solar systems represents a node, and the line of the average distance between each solar system represents an edge.
B. The graphical depiction of the nodes (exoplanets) in the chart, such as their size and color, and the graphical depiction of the edges, such as their thickness and color, would be determined by planetary-scale measures associated with the extent to which conditions in space facilitate a potential ET communication network. A planetary-scale measure as such that can determine the graphical depiction of nodes can include, though not necessarily be limited to, a planet's diameter, mass, or even a composite/index measure. And a planetary-scale measure that can determine the graphical depiction of edges can include the degree of space dust between each neighboring exoplanet, the orbital eccentricities of each planet around its host star(s), or a composite/index measure.
C. After applying the aforementioned data visualization and network analysis, the LPNs and/or WPNs where potential ET communications networks would most likely be (based on the planetary-scale measures for nodes and edges previously described) would graphically stand out in the 3D astronomical chart via highlighted nodes and edges.

A simplified example of this in terms of network analysis would be to consider three neighboring exoplanets called Alpha, Beta, and Gamma in a potentially habitable solar system. Consider that exoplanet Alpha would be a gas giant like Jupiter, while exoplanets Beta and Gamma would be solid Earth-like planets. If we graphically depicted solid Earth-like planets as nodes represented by solid black circles, and gas giants as nodes represented by hollow black circles, then Alpha, Beta, and Gamma would be depicted accordingly. Now consider that there would be an observable dust cloud or similar barrier between Alpha and Beta (but not between Beta and Gamma), which could impair the ability of a potential ET communication network to span between Alpha and Beta, particularly where such a network entailed radio transmissions. If we graphically depicted edges in bold between neighboring nodes for when there would be no dust cloud or similar communication barrier in between the nodes, and unbold edges for barrier conditions, then the edges between the nodes (representing Alpha, Beta, and Gamma) would be depicted accordingly. In sum, in this simplified example, planetary networks depicted mainly by solid black circles connected with bold edges can highlight where potential ET communication networks would likely be.

Considering the aforementioned discussion, the two hypothesis are as follows, denoted as H1 and H2, respectively:

(H1) An ET technosignature would more likely appear in a potentially habitable solar system containing a high concentration of exoplanets, where the exoplanets function as communication access points to facilitate a potential ET communication network.

(H2) An ET technosignature would more likely appear in a highly concentrated cluster of potentially habitable solar systems, wherein the potentially habitable solar systems (by way of their respective exoplanets) function as communication access points to facilitate a potential ET communication network.

The hypotheses have been inferred from an interdisciplinary approach incorporating concepts from information science, the social sciences, and the physical sciences. These concepts are elaborated on in the following section. Each hypothesis by itself is deemed as an essential, yet not sufficient condition in the likelihood of detecting a technosignature of a communicative ET intelligence.

4.2 Concepts

The hypotheses incorporate the concepts of communicative ET intelligence, potentially habitable solar systems, and space-based ET communications networks. Each of these concepts is described as follows, to provide context relative to the hypotheses.

Communicative ET Intelligence

A communicative ET intelligence is a species "not originating recently on Earth" that transmits technosignatures through space consisting of information on carriers such as photons [4.21] [4.22]. A communicative ET intelligence would be represented by a factor in the renowned Drake equation, shown below as Equation 4.1:

$$N = R_{*} f_p \, n_e f_l f_i f_c \, L \tag{4.1}$$

In the equation, N represents the number of communicative intelligent civilizations in the Milky Way galaxy that are detectable. R_{*} represents the average rate of star formation in the Milky Way. f_p is the fraction of stars in the Milky Way that have planets. n_e is the average number of planets in the Milky Way that support life for each star that has planets. f_l represents the fraction of planets in the Milky Way that can support life and where life would actually develop. f_i represents the fraction of planets in the Milky Way that have life that develops into intelligent life. f_c is the fraction of such intelligent life that transmits detectable signals into space. And L is the length of time for which such intelligent life transmits detectable signals into space [4.4].

Considering the virtually ubiquitous evolutionary processes on Earth, such as the biological [4.2], sociological/anthropological [4.12], and informational [4.20], it is conceivable that similar evolutionary processes can occur on a potentially habitable exoplanet such that a communicative ET intelligence can emerge [4.5], notwithstanding that it would not necessarily be in lockstep with our evolution. It can be as though evolution in general is a cosmic universal, analogous to a cultural universal such as music found throughout several regional human societies here on Earth [4.1].

Potentially Habitable Solar Systems

A potentially habitable solar system, beyond Earth's own solar system, is a solar system that contains at least one exoplanet potentially located in the solar system's habitable zone. The habitable zone is a circumstellar region within a solar system wherein a terrestrial mass planet, having between 0.1 and 10 Earth masses and favorable atmospheric conditions, can sustain liquid water on its surface [4.6] [4.7] [4.8] [4.9] [4.10] [4.16] [4.19]. Such a planet can support life ranging from microorganisms to plants and animals (one reference for determining potentially habitable solar systems involving exoplanets is the Habitable Exoplanets Catalog maintained by the Planetary Habitability Laboratory at the University of Puerto Rico at Arecibo). A potentially habitable solar system does not mean that life necessarily exists there, rather it has the potential to exist [4.15]. There can be an incredible opportunity to discover potentially habitable solar systems beyond Earth, considering that there can be as many as 11 billion Earth-size exoplanets orbiting the habitable zone of sun-like stars in the Milky Way galaxy [4.14].

Hypothetical ET Communication Networks in Space

The network analysis approach in conjunction with the hypotheses are based on the following idea: the limits of intelligent communication across space can be captured by supposing the existence of communication networks. While we are only familiar with our own modern space-based communication network extending from Earth, we can draw upon its characteristics as a basic analogy to aid us in locating where in space a potential ET communication would most likely be. The physical topology of our space-based communication network effectively uses the planets in our own solar system as nodes, and the lines of average distance (mean communication paths) between adjacent planets as edges. For instance, consider the planet Jupiter having served as node to help facilitate NASA's space-based communication network. This significantly took shape in 1973 when NASA's Pioneer 10 space probe reached Jupiter. Upon reaching Jupiter, Pioneer 10 carried out a multitude of experiments ranging from studying interplanetary and planetary magnetic fields, solar wind parameters, cosmic rays, to the atmosphere of Jupiter and some of its satellites. Pioneer transmitted data from these experiments back to Earth, in the megabytes range. This contributed to our ability to use this information, and Jupiter's gravitational pull, to network further on to Saturn, Uranus, and Neptune with the subsequent Voyager 1 and 2 missions several years later [4.11]. Moreover, by having a higher concentration of planets in our solar system (eight planets versus less than that), conditions have been more favorable for us to facilitate a communication network in space.

The second hypothesis pertaining to the likelihood of finding ET technosignatures in a highly concentrated cluster of potentially habitable solar systems considers the following: If a communicative ET intelligence has the capability to expand its space-based communications network beyond its own solar system to another one, it can do so more conveniently if the other solar system was close by instead of far away, especially if there would be considerations such as potential attenuation of high-frequency components of an ET transmission technosignature from interstellar scattering between solar systems. It "will require special transmission schemes if broadcasts are made over long distances in the galactic plane" [4.18]. This is with respect to Shannon's Law, shown below as Equation 4.2, which determines "the amount of information C (in bits/second) that can be conveyed with a channel of bandwidth W" and a signal-to-noise ratio of S/N [4.17].

$$C = W log_2\left(1 + \frac{S}{N}\right) \tag{4.2}$$

4.3 Case Study

One example of an application to determine potential ET communication networks in space is a network analysis of 96 exoplanets targeted for observation by the French radio telescope NenuFAR during 2019. The analysis

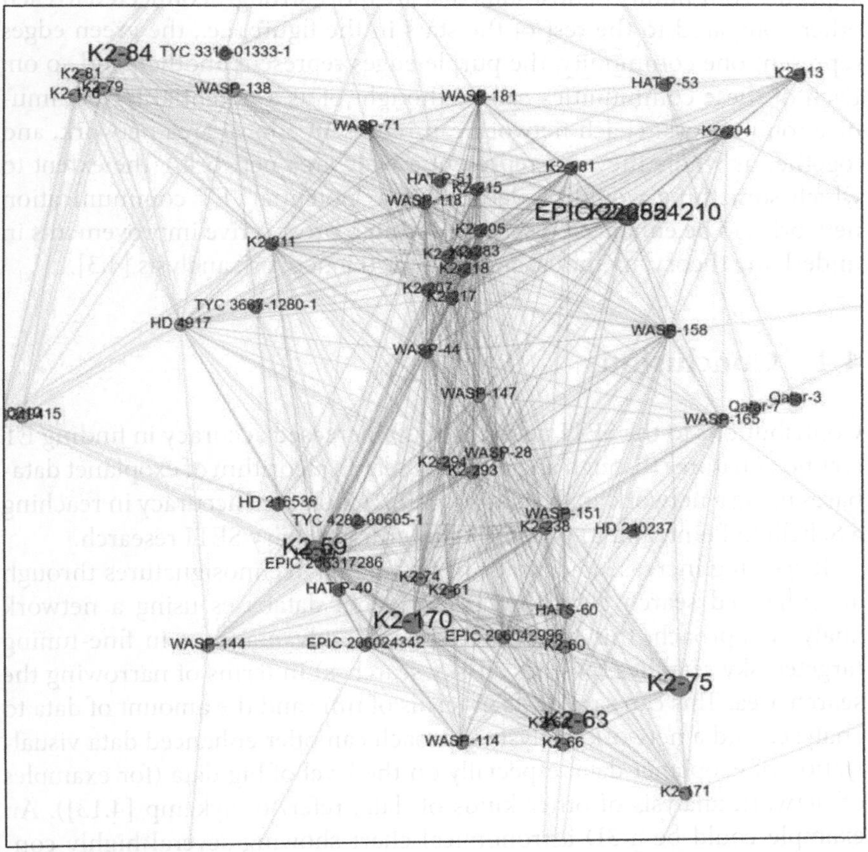

Figure 4.1 A visualization of modular network analysis of 96 exoplanetary systems targeted for SETI observation using the NenuFAR radio telescope. (Figure courtesy of Caleb Jones).

was performed using a Python data science platform in conjunction with data from the NASA Exoplanet Archive online. Figure 4.1 is a visualization of this network analysis (the visualization is part of a larger visualization not shown here).

In the figure, each exoplanet system's host star represents a node in the network, while a hypothetical communication or transport path represents an edge (or line). The size of each node is determined by the number of confirmed or candidate exoplanets in the system: the more exoplanets there are in a system, the larger its node. The weight and length of each edge is determined using a single connected component set to a 300 parsec edge "max" distance, in conjunction with a type of network analysis known as modularity analysis. Edges of a specific color are associated with a specific community of host stars that are more strongly connected to each other compared to the rest of the stars in the figure, i.e., the green edges represent one community, the purple edges represent another, and so on. Each of these communities can be thought of as a potential ET communication network. (Each network can represent a local area network, and together as whole they can represent a wide area network.) The extent to which something would be deemed as a "potential" ET communication network can be enhanced over time through prospective improvements in underlying theory, methods, and data for the network analysis [4.3].

4.4 Conclusions

Contributions to the SETI field can be 1) increased accuracy in finding ET technosignatures through an enhanced search algorithm of exoplanet databases using a network analysis approach, 2) increased accuracy in reaching a Schelling Point, and 3) promoting interdisciplinary SETI research.

Regarding increased accuracy in finding ET technosignatures through an enhanced search algorithm of exoplanet databases using a network analysis approach, this would be especially advantageous in fine-tuning targeted sky searches (besides all-sky searches) in terms of narrowing the search area. This can save costs in terms of time and the amount of data to analyze. And a network analysis approach can offer enhanced data visualization of exoplanet data, especially on the level of big data (for examples of network analysis of other kinds of data, refer to Nykamp [4.13]). An example could be a 3D astronomical chart showing several highly concentrated clusters of potentially habitable solar systems. The solar systems in each cluster would be depicted as nodes, and the distance lines (a communication path) between each solar system depicted as edges. The nodes and edges in each cluster would represent a potential ET communication networks. The chart can also show potentially habitable solar systems not considered part of a cluster, yet where these solar systems would still have a high concentration of planets.

As far as increased accuracy in reaching a Schelling Point, this is the point at which we and a communicative ET intelligence both happen to come to the same realization of how we both should try to discover each other, prior to first contact. An example is a communicative ET intelligence setting up a beacon to transmit a radio signal on a "magic frequency"—the frequency that the ET intelligence hopes we will dial into; the signal may say something as simple as "We are here!" [4.21]. Besides searching for the magic frequency, SETI researchers can draw upon the hypotheses to also consider where such a frequency may likely originate from in the cosmos. This can help narrow the search for such a frequency, thus saving time, money, and other resources in the process.

Since the hypotheses draw upon concepts from an interdisciplinary perspective (such as network analysis used in information science, the social sciences, and the physical sciences), it can promote interdisciplinary research.

A limitation of the hypotheses is that their rationale would mainly apply to communicative ET intelligences whose evolution would be basically comparable with ours here on Earth. Thus, this may be interpreted as being anthropocentric to an extent. Yet the approach does not presume that the evolution of a communicative ET intelligence would necessarily parallel ours in every aspect. Nevertheless, it is conceivable that among the estimated 11 billion Earth-like exoplanets that can be in our galaxy, there can be a chance for a communicative ET intelligence whose evolution is comparable to ours.

Acknowledgements

Thanks to the Lunar and Planetary Institute, Universities Space Research Association, and NASA for organizing the 2018 NASA Technosignatures Workshop that helped to inspire this chapter. Thanks to Seth Shostak of the SETI Institute, David Grinspoon of the Planetary Science Institute, and Kathryn Denning of York University for suggesting sources, and Ronaldo Vigo of Ohio University for providing editorial assistance. And thanks to other SETI colleagues who responded to related scientific questions via the International Academy of Astronautics (IAA) SETI Committee's Google group forum.

References

[4.1] Brown, D.E. (1991) *Human Universals.* Temple University Press, Philadelphia.

[4.2] Darwin, C.R. (1876) *The Origin of Species by Means of Natural Selection, or the Preservation of Favoured Races in the Struggle for Life*, 6th edition. John Murray, London.

[4.3] Davis, R, Jones, C., and Hellbourg, G. (2019) Network analysis of exoplanets for NenuFAR SETI. http://calebjones.info/seti/nenufar-analysis/index.html

[4.4] Drake, F. (1965) The radio search for intelligent extraterrestrial life. In *Current Aspects of Exobiology*. G. Mamikunian and M.H. Briggs (eds.), Oxford University Press: 323–345.

[4.5] Grinspoon, D. (2018) Cognitive planetary transitions: An astrobiological perspective on the "Sapiezoic Eon". Presented at the 2018 NASA Technosignatures Workshop, Lunar and Planetary Institute, Houston, TX. https://www.hou.usra.edu/meetings/technosignatures2018/presentation/?video=grinspoon.mp4

[4.6] Hart, M.H. (1978) The evolution of the atmosphere of the earth. *Icarus* 33, 23-39.

[4.7] Huang, S.S. (1959) Occurrence of life outside the solar system. *American Scientist* 47, 397.

[4.8] Kaltenegger, L. and Sasselov, D. (2011) Exploring the habitable zone for Kepler planetary candidates. *The Astrophysical Journal Letters* 736(2), L25.

[4.9] Kasting, J.F., Whitmire, D.P., and Reynolds, R.T. (1993) Habitable zones around main sequence stars. *Icarus* 101(1), 108–228.

[4.10] Kopparapu, R.K., Ramirez, R., Kasting, J.F. Eymet, V., Robinson, T.D., Mahadevan, S.,…Deshpande, R. (2013) Habitable zones around main-sequence stars: New estimates. *The Astrophyical Journal* 765, 131.

[4.11] NASA. (2018) Pioneer 10. NASA Space Science Data Coordinated Archive (NSSDCA). https://nssdc.gsfc.nasa.gov/nmc/spacecraft/display.action?id=1972-012A

[4.12] Nolan, P. and Lenski, G. (2015) *Human Societies: An Introduction to Macrosociology*. Oxford University Press, New York, NY.

[4.13] Nykamp D.Q. (2019) An introduction to networks. http://mathinsight.org/network_introduction

[4.14] Patigura, E.A., Howard, A.W., and Marcy, G.W. (2013) Prevalence of Earth-size planets orbiting Sun-like stars. Proceedings of the National Academy of Sciences, November 2013. https://www.pnas.org/content/pnas/110/48/19273.full.pdf

[4.15] Planetary Habitability Laboratory, University of Puerto Rico at Arecibo. (2018) HEC: About the Habitable Exoplanets Catalog. http://phl.upr.edu/projects/habitable-exoplanets-catalog/about

[4.16] Selsis, F., Kasting, J.F., Levrard, B., Paillet, J., Ribas, I., and Delfosse, X. (2007) Habitable planets around the star Gliese 581? *Astronomy and Astrophysics* 476(3), 1373–1387.

[4.17] Shannon, C.E. (1948) A mathematical theory of communication. *Bell System Technical Journal* 27, 379-423, 623–656.

[4.18] Shostak, G.S. (1995) SETI at wider bandwidths? In *Progress in the Search of Extraterrestrial Life: 1993 Bioastronomy Symposium*, 74, G.S. Shostak (ed.), Astronomical Society of the Pacific, San Francisco: 447–454.

[4.19] Underwood, D.R., Jones, B.W., and Sleep, P.N. (2003) The evolution of habitable zones during stellar lifetimes and its implications on the search for extraterrestrial life. *International Journal of Astrobiology* 2(4), 289–299.

[4.20] Vigo, R. (2013) Complexity over uncertainty in generalized representational information theory (GRIT): A structure-sensitive general theory of information. *Information* 4, 1–30.

[4.21] Wright, J.T. (2018) Taxonomy and jargon in SETI as an interdisciplinary field of study. From the 2018 Decoding Alien Intelligence Workshop, SETI Institute, Mountain View, CA. https://daiworkshop.seti.org/sites/default/files/workshop-2018/Wright%20-%20Taxonomy%20and%20Jargon%20in%20SETI%20as%20an%20Interdisciplinary%20Field%20of%20Study.pdf

[4.22] Wright, J.T., Sheikh, S., Almár, I., Denning, K., Dick, S., and Tarter, J. (2018) Recommendations from the ad hoc committee on SETI nomenclature. https://arxiv.org/ftp/arxiv/papers/1809/1809.06857.pdf

5

Habitable Mini-Earths with Black Hole Cores

Branislav Vukotić[1]* and Richard Gordon[2,3†]

[1]Astronomical Observatory, Belgrade, Serbia
[2]Gulf Specimen Marine Laboratory & Aquarium Panacea, FL, USA
[3]C.S. Mott Center for Human Growth & Development, Department of Obstetrics & Gynecology, Wayne State University, Detroit, MI, USA

Abstract

Those contending for the planned colonization of our solar system have two basic concepts: terraforming Mars or constructing habitable spinning hollow cylinders in near Earth orbits. Mars is further from the sun and therefore quite cold at its surface. Cylinders are subject to high radiation and puncture. We propose the third possibility of habitable mini-Earths that have a black hole core of sufficient size to retain an atmosphere and Earth-like gravity at the surface. The results of our investigation suggest that the ability of a mini-Earth to have Earth-like atmospheric conditions on the surface strongly depend on mass and size of such a habitat. The Earth-like surface conditions can be achieved in small-size habitats, with radius of <100 kilometers and with the relatively small mass black hole cores of $\sim 10^{20}$ kg. Mass boosting also offers prospects for terraforming of existing objects, with sizes ranging within magnitude orders of $10\text{-}10^3$ km. Future searches for biosignatures on Earth-like exoplanets might be able to detect Earth-like atmospheres on such objects.

Keywords: Mini-Earths, black holes, technosignatures, biosignatures, space colonies, interstellar travel

**Corresponding author*: bvukotic@aob.rs
†*Corresponding author*: DickGordonCan@protonmail.com

Anamaria Berea (ed.) Technosignatures for Detecting Intelligent Life in Our Universe: A Research Companion, (69–84) © 2022 Scrivener Publishing LLC

5.1 Introduction

At the beginning of the twenty-first century, the rapid discoveries of exoplanets highlighted the contradiction of the apparent SETI failure. It was evident that the traditional SETI approaches needed a serious upgrade in order to keep up with the growing confidence that, at least, simple life is widespread in the Milky Way. In a matter of two decades, SETI research has advanced from opportunistic undertakings to modern strategies that rely on the latest scientific achievements across a multitude of relevant research fields [5.16]. With the principles of Copernicanism [5.24], the constraints resulting from SETI efforts represent the unique opportunity to project the future of humanity and its role in the galaxy. The modern SETI thus forms a unified research frame together with the future of humanity studies.

In this work we argue the possibility of making a multitude of habitats smaller than Earth, in orbit around the Earth or the sun, within the classical habitable zone (cf.[5.7]), but with the similar level of comfort of climates in Earth, in order to facilitate the spreading of humanity across the solar system and at least our galaxy. Our primary goal is to underline the most optimal parameters for this kind of habitat in terms of composition and size, given the available amount of building material. Our present knowledge of solar system planets, and also extra-solar worlds, suggests that the mass of such objects stands out as the basic factor that defines the overall habitability characteristics. While small habitats are more feasible to build in larger numbers, the larger ones might have more comfortable Earth-like surface conditions.

Bayesian analysis of [5.4] [5.7] suggests that the long-term survival of humanity rather indicates the large probability of an upcoming catastrophe and not the small probability of such events so far over the entire life span of human civilization. The future of humanity is more likely to be secured on a multitude of habitats rather than a single one on Earth. One of the possible benefits of the Blue Origin [5.3] and pressurized shell worlds concepts [5.17], is that the large number of self-sufficient space colonies would significantly increase the survival probability against naturally induced hazards, such as secular changes in solar activity, stellar collisions, asteroid impacts, etc. However, the suggested hollow cylinders or spherical shells might be sensitive to puncture or damage by high-speed small asteroids or meteors. The mini-Earths approach might offer better protection from such events with the Earth-like atmosphere at its surface acting also as a shield.

5.2 Surface Comfort

It is hard to conceive the lush biosphere on Earth and diversity of plant and animal life without a sustainable layer of atmosphere. Living forms incorporation of non-living matter into their cells and tissues to sustain life requires constant cycling of matter between the biosphere and its host planetary body. Together with other aerobic organisms in the Earth's biosphere, humans strongly depend on availability of oxygen and its replenishment through feedback cycles that happen in the interaction between atmosphere and the surface layer of the Earth [5.10]. When compared to the need for oxygen (in its optimal ratio with other constituents of air), humans can spend significantly more time without water and food. The ability to move and breath freely is essential for the optimal functioning of our nervous systems that have evolved under the surface conditions on Earth. Our evolution has deeply rooted us into the interaction between the atmosphere and surface of the Earth. In order to account for the best living conditions for humans over longer time periods, the designs of various space-born habitats and vehicles attempt to imitate the surface conditions on Earth, as close as possible under the circumstances of sustaining a fragile human life in the harsh conditions of open space. The closer this imitation gets to the real surface conditions in terms of gravity, pressure and air composition, the higher the level of comfort is achieved and humans can function up to their full potential that they have evolved during their evolution on the surface of Earth.

Besides the comfort, a successful space-born vehicle (colony) should be able to sustain itself over long periods of time. The sustainability of Earth is obvious and designing the imitation of such a system is one way to extrapolate our knowledge on space-born habitats. In order to achieve the optimal building costs, we argue the possibility of making Earth-like habitats smaller, dubbed mini-Earths. However, smaller habitats of similar surface composition would likely have smaller surface gravity. To achieve the Earth-like level of surface gravity at the smallest sizes, we investigate the possibility to boost the mini-Earth mass by incorporating mini black holes into their design, as the black holes are the most compact state of matter in the universe.

In the next section we present a concept of the mini-Earths with black hole cores. The optimal range of masses and sizes for such habitats, in order to maintain the Earth-like surface conditions is discussed.

5.3 Concept and Design

The basic idea of the mini-Earth concept is to find a set of basic parameters that define a habitat in such a way that those habitats can be produced in large numbers, but at the same time to keep the comfort level as close as possible to the one that exists on Earth, with the atmosphere similar to the one on Earth. From the present knowledge on exo-worlds, both, in and out of the solar system, the mass of the habitat is the basic parameter. The basic classification of exoplanets is performed by mass. In a similar way, the mass influences the orbital properties and also influences all other parameters to achieve the desired level of comfort.

5.4 Surface Conditions and Size

We have performed a simple calculation of the surface habitability conditions of a body based on its mass. Surface gravity is calculated as:

$$\gamma \frac{m}{r^2}$$

where γ is the gravitational constant $6.67 \times 10^{11} m^3 kg^{-1} s^{-2}$, m is the mass of the body and r is the radius of the body. The hydrostatic pressure (p) equation is applied in isothermal ideal gas approximation, to calculate the mass of the planetary body:

$$\int_{p_s}^{p_e} \frac{dp}{p} = \frac{-\gamma M}{RT} m \int_0^{h_e} \frac{dh}{r(r+h)^2},$$

$$m = \frac{-RT}{\gamma M} ln\left(\frac{p_e}{P_s}\right) \frac{r(r+h_e)}{h_e},$$

where, p_s =101.3 kPa is the surface pressure on Earth, p_e = 26.5 kPa is the atmospheric pressure at the top of the troposphere at h_e = 10km, above the sea level [5.14]. The universal gas constant is R = 8.314 J/(mol K), while M = 0.02897 kg/mol is the molar mass of dry air. Although the air temperature decreases with height approximately as 6.5 K/km [5.14], for the sake of simplicity, we have assumed a constant air temperature of T = 273 K.

A simple model for Earth's atmosphere usually assumes that gravitational acceleration is constant with height, since the atmospheric scale height is several orders of magnitude smaller than the size of the Earth. As this work investigates a possibility for shrinking to smaller sizes, we have adopted the $g(h) \sim (r+h)^{-2}$ dependence.

It is evident from the equations that the pressure height scale is strongly dependent on m and r, i.e., surface gravity and $g(h)$ and change in these factors is likely to significantly influence the surface pressure. Since the troposphere is the lowest and the densest layer of Earth's atmosphere, it contains around ¾ of the total atmospheric mass and our model will approximate the height of the troposphere as the height of the atmosphere. In Figure 5.1, we plot the required body mass ranges in order to satisfy the model conditions for the height of the atmosphere and surface radius of the body.

It is evident that if the radius of the body is two orders of magnitude smaller than the radius of the Earth, it is still possible to have Earth-like surface conditions in terms of similar surface gravity and pressure. This is clearly achieved with mass of the planet being lowered to $\sim 10^{20}$ kg, four orders of magnitude smaller than the mass of the Earth, in order to keep the

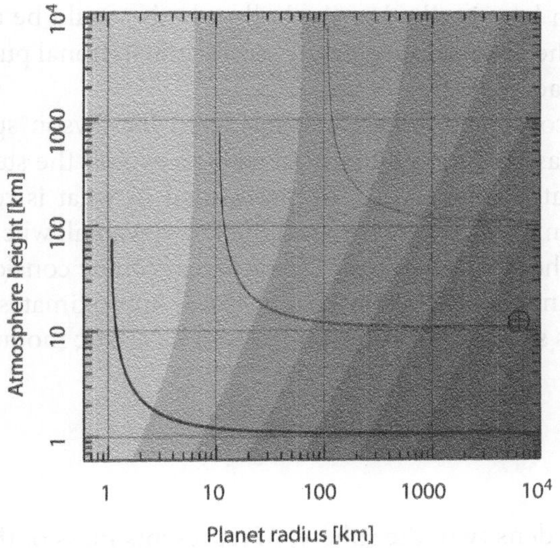

Figure 5.1 Atmospheric surface conditions for various sizes of the planetary body. Gray-scale bins signify the body mass order of magnitude, with lighter shades representing smaller masses and vice versa. The position of the Earth is presented with a symbol in the $\sim 10^{24}$ kg bin. The thin line designates the surface gravity of 0.1g, while the thick line presents the 10g, with Earth being approximately on the 1g line. $g = 9.81$ m/s^2.

similar surface gravity at the smaller surface radius. Assuming a constant density of the planet's material and volume scaling as r^3, the size reduction of two orders of magnitude produces the six orders of magnitude reduction in mass. From Figure 5.1, an Earth-like surface condition at 60km radius could then be achieved for significant higher body densities. For a 60 km radius, the density required to achieve $\sim 10^{20}$ kg mass amounts to $\sim 10^6$ kg/m^3, which is not attainable in the usual states of matter. A possible way to achieve the higher body mass is to boost it by a compact massive body of small size and extremely high density, such as a black hole. Although white dwarfs and neutron stars also have densities that are much higher than $\sim 10^6$ kg/m^3 [5.12] they are not nearly as compact as black holes [5.5]. Since they are not singularity objects, reducing their size would reduce the gravitational attraction and they would disperse to smaller densities and larger sizes. A typical neutron star is about 10 km in radius, while the white dwarfs are significantly larger [5.12]. Also, the neutron stars could only be captured or made with the stellar mass amounts of matter, while black holes could be made with much smaller amount of matter (energy) by achieving an *in situ* singularity.

When compared to the 60 km radius, a neutron star would need a significant size of the cavity to be incorporated into the mini-Earth. The major concern here is whether the hollow body would be able to sustain itself against the collapse, despite the strong gravitational pull of the incorporated compact mass.

A similar construction case is considered for Dyson spheres [5.6], a shell-like megastructure built around stars to exploit the stellar energy, in [5.23]. The stated calculations are performed in what is readily called a "thin" approximation, where the shell thickness is small when compared to the radius of the shell so that the elastic forces can be considered constant over the thickness of the shell. Wright [5.23] approximates that a monolithic sphere is susceptible to buckling given its elastic modulus is:

$$E \approx \frac{\gamma \rho M}{t},$$

where ρ is the density of the sphere, M represents mass of the body inside the sphere and t is the shell thickness. Using values that are indicative for materials readily found in interplanetary space, such as ice, stone or metals, $\rho = 10^3$ kg/m^3 and $E = 100$ GPa, the minimal shell thickness of such a structure amounts to $t \sim 10$ m. Building an $r = 60$ km body of 10 m thickness would roughly require a $\sim 10^{11}$ m^3 of material which is roughly the amount of concrete required to build a highway around the Earth a thousand times. However, this does not appear costly when viewed as a space habitat with a Belgium-sized surface area. Clearly, there is more than enough building material in the solar system to build such mini-Earth shells in large numbers.

5.5 Black Hole Core

Although the standard mass black holes are found in the central part of galaxies, recent work [5.20] announces the indications of stellar-mass black holes in the core of a star cluster. They suggest that these black holes might be the stellar remnants that sank to the cluster core due to dynamical friction. Extrapolating this to find a sub-planetary-mass black holes does not seem justified since gravitational collapse can only make black holes out of stellar mass objects. However, the Universe surprised us with many discoveries so far and this might point us in the direction of eventually finding miniature black holes, or understanding the related phenomena in order to make one using technology treats of the future. Apart from constructing a shell and atmosphere, the building prospects for mini-Earths are likely to depend much more on the possibility to build a singularity, either from matter or from energy. Energetic equivalent of 10^{20} kg of mass is of the order of a thousand years of total sun's luminosity. While such energy can be harvested with Dyson spheres, developing a storage technology for that amount of energy, before using it to initiate a singularity phase transition, has yet to be envisioned.

Another building approach that can be considered is to transform the existing states of matter into black holes, possibly by initiating physical processes that are on the level of a complete transformation of chunk of matter into energy. This would require understanding of physical laws on a level compared to our historical development of energy release, from mastering fire to nuclear reactors. Crane & Westmoreland [5.5] have discussed that a ~10^9 kg black holes might be built by converging lasers. If the lasers are focused to the center of the spherical solid body of a mere 300 km in radius, it could initiate a controlled collapse of such a sphere into the 10^{20} kg black hole.

The black holes with 10^{20} kg masses have evaporation times of ~10^{34} yr, which makes them extremely stable. In addition, the Schwarzschild radius of such a black hole would be of the order of up to a micrometer, which is clearly advantageous from the point of making a black hole confinement structure. One possibility is to use electromagnetic force, from the magnetized shell constructed around the core or on the inside wall of the mini-Earth. For this purpose, charged black holes should be used. According to [5.8], charged black holes might even have larger lifetimes than their charge-less counterparts. Crane & Westmoreland [5.5] imply that initially charged black holes might quickly radiate away most of their charge.

Also, the Hawking luminosity of a black hole of mass m behaves as $\sim m^{-2}10^{34}$, which is of the order of a 10^{-6} W for $m \sim 10^{20}$ kg. This is not nearly sufficient to effectively maneuver or power the mini-Earth, requiring an additional power source. This might be achieved with power generators that use black holes of smaller mass and are positioned on the outside surface of the mini-Earth, or sunlight. A possibility for black hole powered starships is investigated in [5.5] and [5.11].

5.6 Mass Boosting and Terraformation

Part of the 1*g* curve in Figure 5.1, does not depend on radius for sizes greater than ~10 km. This allows a possibility to mass boost the existing objects in size range from 10 km up to the size of the Earth. This possibility was briefly mentioned in [5.2] [5.3]. The inner solar system has numerous objects in this size range including: Mercury, Mars, five dwarf planets, our moon, tens of satellites of giant planets, ~200 asteroids larger than 100 km in diameter [5.9] and many more such bodies that are yet to be discovered. A seemingly inexhaustible reservoir of such objects is likely to be the Oort's Cloud at the outer solar system. Obviously, apart from building shells around black hole cores, there is a significant possibility of boosting the existing objects and making them amenable to terraformation. In addition to considering exoplanet terraforming (e.g. [5.18]), the mass boosting approach opens up various such possibilities within the solar system. In example, the Moon mass of 10^{22} kg would require an order of magnitude boost, since the corresponding 1g mass range is $\sim 10^{23}$ kg for the Moon radius (Figure 5.1).

5.7 Technosignatures

The non-standard appearance of mini-Earths, otherwise not common for small bodies in the exoplanet systems, offers prospects for their detection. Obviously, such a signature would be the observation of spectral lines of atmospheric constituents on the bodies that are too small in size to support any amount of atmosphere observable from afar. Within the biosignatures searches in the near future, humanity will likely be able to observe Earth-sized atmospheres of terrestrial-like exoplanets [5.13] [5.15], which opens up a possibility of observing the atmospheres of mini-Earths.

Further technosignature prospects would rest on the ability to observe small bodies in exoplanetary systems. While the observation of such objects is limited even within the solar system, the collective effects of single or multitude of mini-Earths might stand out as viable future prospects. Besides the dynamical effects of mini-Earth masses on the surrounding populations of small bodies, the crowded parts of the planetary systems that should otherwise be pretty much empty, such as Lagrange points of the planetary system gravitational potential [5.21] or orbits that are perpendicular to the planetary plane. Although the atmosphere of the mini-Earth should offer significant protection from the collisions with small interplanetary debris, similar risks could further be reduced by placing the habitats in parts of the orbital elements parameter space that is rather devoid of mid- and large-size asteroids and comets. This would probably exploit the possibility that mini-Earths adjust their trajectories and maintain themselves in otherwise unstable configurations of orbital elements.

A shape of the object might also indicate its artificial origin. Almost perfectly spherical ~10-100 km size asteroids are not likely to exist in large numbers. A compact spherical object of such size would thus be a possible indication of a mini-Earth.

5.8 Conclusion

We have demonstrated the possibility to build shell-like habitats, mini-Earths of ~100 km radius. In order to achieve an Earth-like atmosphere and surface living conditions, the mass of such habitats is boosted by a centrally confined 10^{20} kg black hole. Our future capability for creating new, or confining the existing black holes, is critical for producing such objects.

With the detection of gravity waves from the fusion of binary black holes [5.1] [5.19], we may start to get an estimate of their availability in our galaxy.

The envisioned habitats can be built in large numbers from the existing amount of building material in the solar system (up to 0.00001 solar masses available from small bodies and dwarf planets [5.22]. This amounts of up to 10 Earth masses. The collective surface area of the resulting mini-Earths could support an order of magnitude increase of the current human population). In addition, the similar effect might be achieved by mass boosting and subsequent terraformation of the existing numerous small bodies in the solar system. The obvious signature of these habitats would be an object with an atmosphere which is too large given the size of the object. Such observations might be within the reach of the coming searches for biosignatures in exoplanet systems. The details of the construction of mini-Earths we leave to the future generations of humanity.

Acknowledgement

BV was funded by the Ministry of Education, Science and Technological Development of the Republic of Serbia through contract number 451-03-68/2022-14/200002.

References

[5.1] Abbott, B.P., Abbott, R., Abbott, T.D., Abernathy, M.R., Acernese, F., Ackley, K., Adams, C., Adams, T., Addesso, P., Adhikari, R.X., Adya, V.B., Affeldt, C., Agathos, M., Agatsuma, K., Aggarwal, N., Aguiar, O.D., Aiello, L., Ain, A., Ajith, P., Allen, B., Allocca, A., Altin, P.A., Anderson, S.B., Anderson, W.G., Arai, K., Arain, M.A., Araya, M.C., Arceneaux, C.C., Areeda, J.S., Arnaud, N., Arun, K.G., Ascenzi, S., Ashton, G., Ast, M., Aston, S.M., Astone, P., Aufmuth, P., Aulbert, C., Babak, S., Bacon, P., Bader, M.K.M., Baker, P.T., Baldaccini, F., Ballardin, G., Ballmer, S.W., Barayoga, J.C., Barclay, S.E., Barish, B.C., Barker, D., Barone, F., Barr, B., Barsotti, L., Barsuglia, M., Barta, D., Bartlett, J., Barton, M.A., Bartos, I., Bassiri, R., Basti, A., Batch, J.C., Baune, C., Bavigadda, V., Bazzan, M., Behnke, B., Bejger, M., Belczynski, C., Bell, A.S., Bell, C.J., Berger, B.K., Bergman, J., Bergmann, G., Berry, C.P.L., Bersanetti, D., Bertolini, A., Betzwieser, J., Bhagwat, S., Bhandare, R., Bilenko, I.A.B., G., Birch, J., Birney, R., Birnholtz, O., Biscans, S., Bisht, A., Bitossi, M., Biwer, C., Bizouard, M.A., Blackburn, J.K., Blair, C.D., Blair, D.G., Blair, R.M., Bloemen, S., Bock, O., Bodiya, T.P., Boer, M., Bogaert, G., Bogan, C., Bohe, A., Bojtos, P., Bond, C., Bondu, F., Bonnand, R., Boom, B.A., Bork, R., Boschi, V., Bose, S., Bouffanais, Y., Bozzi, A., Bradaschia, C., Brady, P.R., Braginsky, V.B., Branchesi, M., Brau, J.E., Briant, T., Brillet, A., Brinkmann, M., Brisson, V., Brockill, P., Brooks, A.F., Brown, D.A., Brown, D.D., Brown, N.M., Buchanan, C.C., Buikema, A., Bulik, T., Bulten, H.J., Buonanno, A., Buskulic, D., Buy, C., Byer, R.L., Cabero, M., Cadonati, L., Cagnoli, G., Cahillane, C., Bustillo, J.C., Callister, T., Calloni, E., Camp,

J.B., Cannon, K.C., Cao, J., Capano, C.D., Capocasa, E., Carbognani, F., Caride, S., Diaz, J.C., Casentini, C., Caudill, S., Cavaglia, M., Cavalier, F., Cavalieri, R., Cella, G., Cepeda, C.B., Baiardi, L.C., Cerretani, G., Cesarini, E., Chakraborty, R., Chalermsongsak, T., Chamberlin, S.J., Chan, M., Chao, S., Charlton, P., Chassande-Mottin, E., Chen, H.Y., Chen, Y., Cheng, C., Chincarini, A., Chiummo, A., Cho, H.S., Cho, M., Chow, J.H., Christensen, N., Chu, Q., Chua, S., Chung, S., Ciani, G., Clara, F., Clark, J.A., Cleva, F., Coccia, E., Cohadon, P.F., Colla, A., Collette, C.G., Cominsky, L., Constancio, M., Conte, A., Conti, L., Cook, D., Corbitt, T.R., Cornish, N., Corsi, A., Cortese, S., Costa, C.A., Coughlin, M.W., Coughlin, S.B., Coulon, J.P., Countryman, S.T., Couvares, P., Cowan, E.E., Coward, D.M., Cowart, M.J., Coyne, D.C., Coyne, R., Craig, K., Creighton, J.D.E., Creighton, T.D., Cripe, J., Crowder, S.G., Cruise, A.M., Cumming, A., Cunningham, L., Cuoco, E., Dal Canton, T., Danilishin, S.L., D'Antonio, S., Danzmann, K., Darman, N.S., Da Silva Costa, C.F., Dattilo, V., Dave, I., Daveloza, H.P., Davier, M., Davies, G.S., Daw, E.J., Day, R., De, S., DeBra, D., Debreczeni, G., Degallaix, J., De Laurentis, M., Deleglise, S., Del Pozzo, W., Denker, T., Dent, T., Dereli, H., Dergachev, V., DeRosa, R.T., De Rosa, R., DeSalvo, R., Dhurandhar, S., Diaz, M.C., Di Fiore, L., Di Giovanni, M., Di Lieto, A., Di Pace, S., Di Palma, I., Di Virgilio, A., Dojcinoski, G., Dolique, V., Donovan, F., Dooley, K.L., Doravari, S., Douglas, R., Downes, T.P., Drago, M., Drever, R.W.P., Driggers, J.C., Du, Z., Ducrot, M., Dwyer, S.E., Edo, T.B., Edwards, M.C., Effler, A., Eggenstein, H.B., Ehrens, P., Eichholz, J., Eikenberry, S.S., Engels, W., Essick, R.C., Etzel, T., Evans, M., Evans, T.M., Everett, R., Factourovich, M., Fafone, V., Fair, H., Fairhurst, S., Fan, X., Fang, Q., Farinon, S., Farr, B., Farr, W.M., Favata, M., Fays, M., Fehrmann, H., Fejer, M.M., Feldbaum, D., Ferrante, I., Ferreira, E.C., Ferrini, F., Fidecaro, F., Finn, L.S., Fiori, I., Fiorucci, D., Fisher, R.P., Flaminio, R., Fletcher, M., Fong, H., Fournier, J.D., Franco, S., Frasca, S., Frasconi, F., Frede, M., Frei, Z., Freise, A., Frey, R., Frey, V., Fricke, T.T., Fritschel, P., Frolov, V.V., Fulda, P., Fyffe, M., Gabbard, H.A.G., Gair, J.R., Gammaitoni, L., Gaonkar, S.G., Garufi, F., Gatto, A., Gaur, G., Gehrels, N., Gemme, G., Gendre, B., Genin, E., Gennai, A., George, J., Gergely, L., Germain, V., Ghosh, A., Ghosh, A., Ghosh, S., Giaime, J.A., Giardina, K.D., Giazotto, A., Gill, K., Glaefke, A., Gleason, J.R., Goetz, E., Goetz, R., Gondan, L., Gonzalez, G., Castro, J.M.G., Gopakumar, A., Gordon, N.A., Gorodetsky, M.L., Gossan, S.E., Gosselin, M., Gouaty, R., Graef, C., Graff, P.B., Granata, M., Grant, A., Gras, S., Gray, C., Greco, G., Green, A.C., Greenhalgh, R.J.S., Groot, P., Grote, H., Grunewald, S., Guidi, G.M., Guo, X., Gupta, A., Gupta, M.K., Gushwa, K.E., Gustafson, E.K., Gustafson, R., Hacker, J.J., Hall, B.R., Hall, E.D., Hammond, G., Haney, M., Hanke, M.M., Hanks, J., Hanna, C., Hannam, M.D., Hanson, J., Hardwick, T., Harms, J., Harry, G.M., Harry, I.W., Hart, M.J., Hartman, M.T., Haster, C.J., Haughian, K., Healy, J., Heefner, J., Heidmann, A., Heintze, M.C., Heinzel,

G., Heitmann, H., Hello, P., Hemming, G., Hendry, M., Heng, I.S., Hennig, J., Heptonstall, A.W., Heurs, M., Hild, S., Hoak, D., Hodge, K.A., Hofman, D., Hollitt, S.E., Holt, K., Holz, D.E., Hopkins, P., Hosken, D.J., Hough, J., Houston, E.A., Howell, E.J., Hu, Y.M., Huang, S., Huerta, E.A., Huet, D., Hughey, B., Husa, S., Huttner, S.H., Huynh-Dinh, T., Idrisy, A., Indik, N., Ingram, D.R., Inta, R., Isa, H.N., Isac, J.M., Isi, M., Islas, G., Isogai, T., Iyer, B.R., Izumi, K., Jacobson, M.B., Jacqmin, T., Jang, H., Jani, K., Jaranowski, P., Jawahar, S., Jimenez-Forteza, F., Johnson, W.W., Johnson-McDaniel, N.K., Jones, D.I., Jones, R., Jonker, R.J.G., Ju, L., Haris, K., Kalaghatgi, C.V., Kalogera, V., Kandhasamy, S., Kang, G., Kanner, J.B., Karki, S., Kasprzack, M., Katsavounidis, E., Katzman, W., Kaufer, S., Kaur, T., Kawabe, K., Kawazoe, F., Kefelian, F., Kehl, M.S., Keitel, D., Kelley, D.B., Kells, W., Kennedy, R., Keppel, D.G., Key, J.S., Khalaidovski, A., Khalili, F.Y., Khan, I., Khan, S., Khan, Z., Khazanov, E.A., Kijbunchoo, N., Kim, C., Kim, J., Kim, K., Kim, N.G., Kim, N., Kim, Y.M., King, E.J., King, P.J., Kinzel, D.L., Kissel, J.S., Kleybolte, L., Klimenko, S., Koehlenbeck, S.M., Kokeyama, K., Koley, S., Kondrashov, V., Kontos, A., Koranda, S., Korobko, M., Korth, W.Z., Kowalska, I., Kozak, D.B., Kringel, V., Krishnan, B., Krolak, A., Krueger, C., Kuehn, G., Kumar, P., Kumar, R., Kuo, L., Kutynia, A., Kwee, P., Lackey, B.D., Landry, M., Lange, J., Lantz, B., Lasky, P.D., Lazzarini, A., Lazzaro, C., Leaci, P., Leavey, S., Lebigot, E.O., Lee, C.H., Lee, H.K., Lee, H.M., Lee, K., Lenon, A., Leonardi, M., Leong, J.R., Leroy, N., Letendre, N., Levin, Y., Levine, B.M., Li, T.G.F., Libson, A., Littenberg, T.B., Lockerbie, N.A., Logue, J., Lombardi, A.L., London, L.T., Lord, J.E., Lorenzini, M., Loriette, V., Lormand, M., Losurdo, G., Lough, J.D., Lousto, C.O., Lovelace, G., Luck, H., Lundgren, A.P., Luo, J., Lynch, R., Ma, Y., MacDonald, T., Machenschalk, B., MacInnis, M., Macleod, D.M., Magana-Sandoval, F., Magee, R.M., Mageswaran, M., Majorana, E., Maksimovic, I., Malvezzi, V., Man, N., Mandel, I., Mandic, V., Mangano, V., Mansell, G.L., Manske, M., Mantovani, M., Marchesoni, F., Marion, F., Marka, S., Marka, Z., Markosyan, A.S., Maros, E., Martelli, F., Martellini, L., Martin, I.W., Martin, R.M., Martynov, D.V., Marx, J.N., Mason, K., Masserot, A., Massinger, T.J., Masso-Reid, M., Matichard, F., Matone, L., Mavalvala, N., Mazumder, N., Mazzolo, G., McCarthy, R., McClelland, D.E., McCormick, S., McGuire, S.C., McIntyre, G., McIver, J., McManus, D.J., McWilliams, S.T., Meacher, D., Meadors, G.D., Meidam, J., Melatos, A., Mendell, G., Mendoza-Gandara, D., Mercer, R.A., Merilh, E., Merzougui, M., Meshkov, S., Messenger, C., Messick, C., Meyers, P.M., Mezzani, F., Miao, H., Michel, C., Middleton, H., Mikhailov, E.E., Milano, L., Miller, J., Millhouse, M., Minenkov, Y., Ming, J., Mirshekari, S., Mishra, C., Mitra, S., Mitrofanov, V.P., Mitselmakher, G., Mittleman, R., Moggi, A., Mohan, M., Mohapatra, S.R.P., Montani, M., Moore, B.C., Moore, C.J., Moraru, D., Moreno, G., Morriss, S.R., Mossavi, K., Mours, B., Mow-Lowry, C.M., Mueller, C.L., Mueller, G., Muir, A.W., Mukherjee, A., Mukherjee, D., Mukherjee, S.,

Mukund, N., Mullavey, A., Munch, J., Murphy, D.J., Murray, P.G., Mytidis, A., Nardecchia, I., Naticchioni, L., Nayak, R.K., Necula, V., Nedkova, K., Nelemans, G., Neri, M., Neunzert, A., Newton, G., Nguyen, T.T., Nielsen, A.B., Nissanke, S., Nitz, A., Nocera, F., Nolting, D., Normandin, M.E.N., Nuttall, L.K., Oberling, J., Ochsner, E., O'Dell, J., Oelker, E., Ogin, G.H., Oh, J.J., Oh, S.H., Ohme, F., Oliver, M., Oppermann, P., Oram, R.J., O'Reilly, B., O'Shaughnessy, R., Ott, C.D., Ottaway, D.J., Ottens, R.S., Overmier, H., Owen, B.J., Pai, A., Pai, S.A., Palamos, J.R., Palashov, O., Palomba, C., Pal-Singh, A., Pan, H., Pan, Y., Pankow, C., Pannarale, F., Pant, B.C., Paoletti, F., Paoli, A., Papa, M.A., Paris, H.R., Parker, W., Pascucci, D., Pasqualetti, A., Passaquieti, R., Passuello, D., Patricelli, B., Patrick, Z., Pearlstone, B.L., Pedraza, M., Pedurand, R., Pekowsky, L., Pele, A., Penn, S., Perreca, A., Pfeiffer, H.P., Phelps, M., Piccinni, O., Pichot, M., Pickenpack, M., Piergiovanni, F., Pierro, V., Pillant, G., Pinard, L., Pinto, I.M., Pitkin, M., Poeld, J.H., Poggiani, R., Popolizio, P., Post, A., Powell, J., Prasad, J., Predoi, V., Premachandra, S.S., Prestegard, T., Price, L.R., Prijatelj, M., Principe, M., Privitera, S., Prix, R., Prodi, G.A., Prokhorov, L., Puncken, O., Punturo, M., Puppo, P., Purrer, M., Qi, H., Qin, J., Quetschke, V., Quintero, E.A., Quitzow-James, R., Raab, F.J., Rabeling, D.S., Radkins, H., Raffai, P., Raja, S., Rakhmanov, M., Ramet, C.R., Rapagnani, P., Raymond, V., Razzano, M., Re, V., Read, J., Reed, C.M., Regimbau, T., Rei, L., Reid, S., Reitze, D.H., Rew, H., Reyes, S.D., Ricci, F., Riles, K., Robertson, N.A., Robie, R., Robinet, F., Rocchi, A., Rolland, L., Rollins, J.G., Roma, V.J., Romano, J.D., Romano, R., Romanov, G., Romie, J.H., Rosinska, D., Rowan, S., Rudiger, A., Ruggi, P., Ryan, K., Sachdev, S., Sadecki, T., Sadeghian, L., Salconi, L., Saleem, M., Salemi, F., Samajdar, A., Sammut, L., Sampson, L.M., Sanchez, E.J., Sandberg, V., Sandeen, B., Sanders, G.H., Sanders, J.R., Sassolas, B., Sathyaprakash, B.S., Saulson, P.R., Sauter, O., Savage, R.L., Sawadsky, A., Schale, P., Schilling, R., Schmidt, J., Schmidt, P., Schnabel, R., Schofield, R.M.S., Schonbeck, A., Schreiber, E., Schuette, D., Schutz, B.F., Scott, J., Scott, S.M., Sellers, D., Sengupta, A.S., Sentenac, D., Sequino, V., Sergeev, A., Serna, G., Setyawati, Y., Sevigny, A., Shaddock, D.A., Shaffer, T., Shah, S., Shahriar, M.S., Shaltev, M., Shao, Z., Shapiro, B., Shawhan, P., Sheperd, A., Shoemaker, D.H., Shoemaker, D.M., Siellez, K., Siemens, X., Sigg, D., Silva, A.D., Simakov, D., Singer, A., Singer, L.P., Singh, A., Singh, R., Singhal, A., Sintes, A.M., Slagmolen, B.J.J., Smith, J.R., Smith, M.R., Smith, N.D., Smith, R.J.E., Son, E.J., Sorazu, B., Sorrentino, F., Souradeep, T., Srivastava, A.K., Staley, A., Steinke, M., Steinlechner, J., Steinlechner, S., Steinmeyer, D., Stephens, B.C., Stevenson, S.P., Stone, R., Strain, K.A., Straniero, N., Stratta, G., Strauss, N.A., Strigin, S., Sturani, R., Stuver, A.L., Summerscales, T.Z., Sun, L., Sutton, P.J., Swinkels, B.L., Szczepanczyk, M.J., Tacca, M., Talukder, D., Tanner, D.B., Tapai, M., Tarabrin, S.P., Taracchini, A., Taylor, R., Theeg, T., Thirugnanasambandam, M.P., Thomas, E.G., Thomas, M.,

Thomas, P., Thorne, K.A., Thorne, K.S., Thrane, E., Tiwari, S., Tiwari, V., Tokmakov, K.V., Tomlinson, C., Tonelli, M., Torres, C.V., Torrie, C.I., Toyra, D., Travasso, F., Traylor, G., Trifiro, D., Tringali, M.C., Trozzo, L., Tse, M., Turconi, M., Tuyenbayev, D., Ugolini, D., Unnikrishnan, C.S., Urban, A.L., Usman, S.A., Vahlbruch, H., Vajente, G., Valdes, G., Vallisneri, M., van Bakel, N., van Beuzekom, M., van den Brand, J.F.J., Van Den Broeck, C., Vander-Hyde, D.C., van der Schaaf, L., van Heijningen, J.V., van Veggel, A.A., Vardaro, M., Vass, S., Vasuth, M., Vaulin, R., Vecchio, A., Vedovato, G., Veitch, J., Veitch, P.J., Venkateswara, K., Verkindt, D., Vetrano, F., Vicere, A., Vinciguerra, S., Vine, D.J., Vinet, J.Y., Vitale, S., Vo, T., Vocca, H., Vorvick, C., Voss, D., Vousden, W.D., Vyatchanin, S.P., Wade, A.R., Wade, L.E., Wade, M., Waldman, S.J., Walker, M., Wallace, L., Walsh, S., Wang, G., Wang, H., Wang, M., Wang, X., Wang, Y., Ward, H., Ward, R.L., Warner, J., Was, M., Weaver, B., Wei, L.W., Weinert, M., Weinstein, A.J., Weiss, R., Welborn, T., Wen, L., Wessels, P., Westphal, T., Wette, K., Whelan, J.T., Whitcomb, S.E., White, D.J., Whiting, B.F., Wiesner, K., Wilkinson, C., Willems, P.A., Williams, L., Williams, R.D., Williamson, A.R., Willis, J.L., Willke, B., Wimmer, M.H., Winkelmann, L., Winkler, W., Wipf, C.C., Wiseman, A.G., Wittel, H., Woan, G., Worden, J., Wright, J.L., Wu, G., Yablon, J., Yakushin, I., Yam, W., Yamamoto, H., Yancey, C.C., Yap, M.J., Yu, H., Yvert, M., Zadrozny, A., Zangrando, L., Zanolin, M., Zendri, J.P., Zevin, M., Zhang, F., Zhang, L., Zhang, M., Zhang, Y., Zhao, C., Zhou, M., Zhou, Z., Zhu, X.J., Zucker, M.E., Zuraw, S.E. and Zweizig, J. (2016) Observation of gravitational waves from a binary black hole merger. *Physical Review Letters* 116(6), #061102.

[5.2] Arthur, I. (2019) Colonizing Black Holes. https://www.youtube.com/watch?v=pxa0IrZCNzg

[5.3] Blue Origin. (2020) Blue Origin. https://www.blueorigin.com/

[5.4] Ćirković, M.M., Sandberg, A. and Bostrom, N. (2010) Anthropic Shadow: Observation Selection Effects and Human Extinction Risks. *Risk Analysis* 30(10), 1495-1506.

[5.5] Crane, L. and Westmoreland, S. (2009) Are black hole starships possible?, https://arxiv.org/abs/0908.1803

[5.6] Dyson, F.J. (1960) Search for artificial stellar sources of infrared radiation. *Science* 131(3414), 1667-1668.

[5.7] Hand, K.P. (2020) *Alien Oceans: The Search for Life in the Depths of Space*. Princeton University Press, Princeton, New Jersey, USA.

[5.8] Hiscock, W.A. and Weems, L.D. (1990) Evolution of charged evaporating black holes. *Physical Review D* 41(4), 1142-1151.

[5.9] Jet Propulsion Laboratory. (2020) JPL Small-Body Database Search Engine. https://ssd.jpl.nasa.gov/sbdb_query.cgi#x

[5.10] Laakso, T.A. and Schrag, D.P. (2017) A theory of atmospheric oxygen. *Geobiology*, 15(3), 366-384.

[5.11] Lee, J.S. (2015) Acceleration of a Schwarzschild Kugelblitz starship. *Journal of the British Interplanetary Society* 68, 105-116.
[5.12] Liebert, J. (1980) White dwarf stars. *Annual Review of Astronomy and Astrophysics* 18, 368-398.
[5.13] Lustig-Yaeger, J., Meadows, V. and Lincowski, A. (2020) The detection and characterization of terrestrial exoplanet atmospheres and the search for habitability and biosignatures. In: *American Astronomical Society Meeting Abstracts*. 230.205.
[5.14] Oceanic, U.S.N., Administration, A. and Force, U.S.A. (1976) *US standard atmosphere, 1976*. National Oceanic and Atmospheric Administration. 76.
[5.15] Quanz, S.P., Absil, O., Angerhausen, D., Benz, W., Bonfils, X., Berger, J.-P., Brogi, M., Cabrera, J., Danchi, W.C. and Defrère, D. (2019) Atmospheric characterization of terrestrial exoplanets in the mid-infared: biosignatures, habitability & diversity. *arXiv:1908.01316*.
[5.16] Reyes, A. and Wright, J.T. (2019) Towards a Comprehensive Bibliography for SETI. *arXiv preprint arXiv:1908.02587*.
[5.17] Roy, K.I., Kennedy, R.G. and Fields, D.E. (2013) Shell worlds. *Acta Astronautica* 82(2), 238-245.
[5.18] Sleator, R. and Smith, N. (2021) Escape from planet Earth: From directed panspermia to terraformation. In: Planet Formation and Panspermia: New Prospects for the Movement of Life through Space [PNSP, Volume in the series Astrobiology Perspectives on Life of the Universe, Series Editors: Richard Gordon & Joseph Seckbach]. B. Vukotić, J. Seckbach and R. Gordon, (eds.) Wiley-Scrivener, Beverly, Massachusetts USA, 119-124.
[5.19] Sutter, P.M. (2020) Are there any black holes left over from the Big Bang?, https://www.space.com/primordial-black-holes-big-bang.html
[5.20] Vitral, E., Mamon and G.A. (2021) Does NGC 6397 contain an intermediate-mass black hole or a more diffuse inner subcluster?. Astronomy & Astrophysics 646, id. A63, 27 pp.
[5.21] Wikipedia. (2020) Lagrangian point. https://en.wikipedia.org/wiki/Lagrangian_point
[5.22] Wikipedia. (2020) Solar System. https://en.wikipedia.org/wiki/Solar_System
[5.23] Wright, J. (2020) Dyson spheres. *Serbian Astronomical Journal*, 200, 1-18.
[5.24] Zimmerman Jones, A. (2019) Copernican Principle. https://www.thoughtco.com/copernican-principle-2699117

6

Technosignatures in Time-Series Photometry

Hector Socas-Navarro[1]*, Shauna Sallmen[2], Eric Korpela[3] and Daniel Angerhausen[4]

[1]*Instituto de Astrofísica de Canarias, Avda Vía Láctea S/N, La Laguna, Tenerife, Spain, Departamento de Astrofísica, Universidad de La Laguna, Tenerife, Spain*
[2]*Physics Department, University of Wisconsin-La Crosse, La Crosse, WI, USA*
[3]*University of California, Berkeley, CA, USA*
[4]*ETH Zurich, Institute for Particle Physics & Astrophysics, Zurich, Switzerland*

Abstract
New opportunities are being enabled by state-of-the-art instrumentation to search for technosignatures (TS) in time-series photometric observations. We review existing proposals for detectable TS around extrasolar planets or their parent stars and discuss them in the context of the "axes of merit". We briefly describe the methods by which such signals could be detected and how challenging that would be with current or upcoming technology.

Keywords: Astrobiology, extraterrestrial intelligence, Planets and satellites: detection, Planets and satellites: atmospheres, Planets and satellites: surfaces, techniques: photometric

6.1 Introduction

The science of exoplanet discovery and characterization has grown tremendously over the last 25 years, becoming one of the major drivers of modern astronomy [6.17]. Sophisticated instrumentation and modeling techniques for the detection of transits in stellar light curves are opening

**Corresponding author*: hsocas@iac.es

Anamaria Berea (ed.) Technosignatures for Detecting Intelligent Life in Our Universe: A Research Companion, (85–102) © 2022 Scrivener Publishing LLC

exciting new windows to search for technosignatures, with new space missions and ground-based observatories starting operations. The promising prospects for even better sensitivity in the near future make this field an attractive opportunity to explore in years to come.

The first suggestion of a technosignature in the scientific literature is that of Dyson spheres [6.4], mega-engineering structures that other civilizations might build to support the largest possible biosphere. Such structures would occult stellar light and produce infrared excess emission. Dyson spheres are colossal structures of star system–size scales. In the 1960s, and for several decades afterwards, this was the minimum size that could be detected with extant instruments.

The large, irregular dimming events discovered in Boyajian's star, unexpected in a star of its type and age [6.3], sparked a renewed interest in the possibility of detecting such stellar system-scale structures or swarms (see e.g., [6.31]).

In just a few decades, as our observing technology has improved, the detectable size for a transiting object has decreased from stellar system scale in the 1970s, to Jupiter-size in the 1990s, and to Earth-size planets in the 2010s. Efforts are currently underway to push this limit further down in order to be able to detect planetary rings or small moons in the near future.

In the search for transiting alien megastructures, we can state a principle of technology balance: as our own technology advances, we are more sensitive to smaller-scale alien technology. This is simply the consequence of being sensitive to more subtle imprints of smaller-scale engineering on the transit light curves. This could be metaphorically illustrated as:

$$O \times T \geq 1$$

where O and T are dimensionless numbers that somehow parameterize the technological capability of the observer and the target civilizations, respectively. Obviously, these parameters are not well defined but nevertheless the expression of this relationship may offer some insights, discussed below. Note that it is assumed that the capabilities to observe and to produce technosignatures both increase in a similar manner with the overall technological development of a civilization. We note elsewhere that increased technological development could increase the ability to *hide* technosignatures from distant observers.

Current human technology does not produce a significant change on the solar light curve. It would be completely undetectable to us if we were observing it from large (interstellar) distances. Essentially, we would not be able to detect our own transit technosignatures from other stars (although this could change in the future, as we discuss below). The fact that we are not able to observe our own technosignatures implies that our current technological level is well below 1 in the technology balance inequality. Accordingly, we would only be able to establish contact with civilizations that are technologically more advanced than ourselves.

The technology balance inequality provides us with an interesting classification of civilizations according to whether their technology parameter is above or below unity. Let us refer to them as "developing" and "advanced" civilizations, respectively. We can see that the first contact of a developing civilization would necessarily have to be with an advanced one. The inequality prohibits a first contact between two developing civilizations. This might be a good thing, since advanced civilizations may be more likely to have passed the so-called "great filter" [6.28] and have learned to coexist with technology.

Transit technosignatures may be short- or long-lived. This is relevant when considering the general problem of synchronicity, i.e., the likelihood that the observer and target civilizations exist at the same time. A long-lived technosignature would alleviate the synchronicity problem but may outlast its creators, marking the location where a civilization once existed but is not necessarily there now. A short-lived technosignature, on the other hand, is more likely to give away the presence of a currently active civilization but is less likely to coexist with us at the same time.

In this chapter, we describe light-curve technosignatures that have been explored in the scientific literature, review factors to consider when designing a search strategy for these technosignatures, and discuss issues related to their detectability.

6.2 Types of Technosignatures

6.2.1 Dyson Spheres

The first suggestion of a technosignature detectable in photometric observations was put forward by Dyson ([6.4], see also the recent review in [6.30]). He envisioned a type of megastructure, currently referred to as Dyson spheres (or swarms), that might be built by civilizations with enormous engineering capabilities. Dyson spheres are often portrayed in popular culture as a discrete structure built with the purpose of collecting energy from the parent star to supply a civilization's needs. However, this picture does not accurately reflect the concept originally described by Dyson in his paper.

The main argument in the 1960 paper was related to population growth and the conversion of energy into waste heat. Ultimately, the size of a population will be limited by the amount of habitable surface in its biosphere. The optimal arrangement of a planetary system to accommodate the largest possible number of individuals would be one in which the planets are dismantled and their matter rearranged into a large sphere. The inner surface would receive energy from the star and act as a habitable biosphere. Dyson found it remarkable that, for the solar system, the radius of such a sphere would be of a few astronomical units (AU), considering typical densities and a shell thickness of a few meters.

The original Dyson spheres were "dark objects", with a hidden star inside the shell, but radiating copiously in the infrared at wavelengths of a few microns. He proposed to conduct a search for infrared point sources, in parallel to the ongoing projects for radio searches, as a way to look for very advanced civilizations in the galaxy.

In a later issue of the same journal, Dyson [6.5] responded to some letters to the editor addressing some concerns about mechanical and dynamical stability of such a megastructure. He argued that the spheres would not be a monolithic structure, but rather a swarm of smaller fragments moving in individual orbits. Starlight would occasionally escape through temporary gaps between the fragments, which might be observable as a dim point source with a pattern of sudden, possibly irregular, occultations.

6.2.2 Transit Beacons

Arnold [6.1] suggested that civilizations might signal their existence by building artificial planet-sized structures with shapes designed to create unnatural transit light curves. A single triangle, two-screen object, or louver-like six-screen object, each with a cross-section similar to that of a Jupiter-sized planet orbiting 1 AU from a sun-like star, would create a transit light curve that differs from a best-fit planetary transit at the level of $\sim 10^{-4}$. These are at the level of detectability with *Kepler*. Although some of the transit residuals are quite similar to those of a ringed planet, this confusion could be removed if the civilization placed groups of such objects using prime numbers or powers of two to attract the attention of a monitoring civilization. In addition to using opaque structures, a civilization could encode information by adding spectral features, for example by utilizing semi-transparent multilayers to create one or more absorption features.

6.2.3 Orbital Mirrors

Korpela *et al.* [6.14] proposed an engineering project of similar scale, but with a useful function besides communication. Planets in the habitable

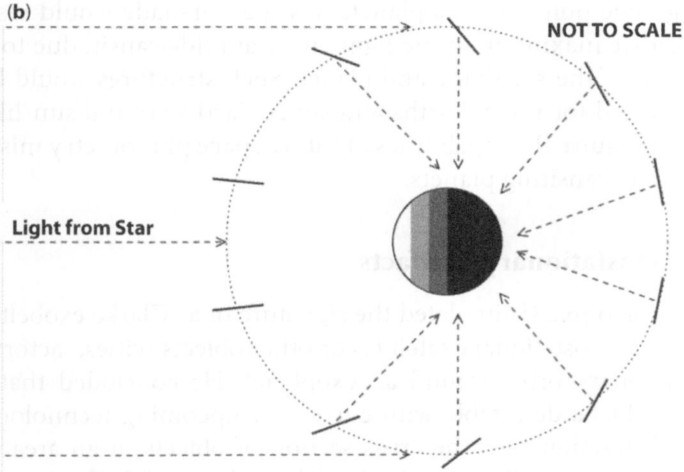

Figure 6.1 Figure 1b from [6.14]. Schematic illustration of how a fleet of orbiting mirrors could redirect starlight onto an exoplanet.

zones of M stars have short tidal-locking timescales, so any old enough to host an advanced civilization in the system are likely to be synchronously rotating. An alien civilization might wish to illuminate the perpetual dark side of their planet, or another planet in the system. One motivation might be to affect the exoplanet's climate.

Although conceptually simple, a single large ($R \sim R_{\text{planet}}$) mirror at the L2 Lagrange point would pose greater engineering challenges than a fleet of numerous independently steerable smaller (1 km^2, 1000 kg) mirrors orbiting the planet (shown schematically in Figure 6.1). Korpela *et al.* [6.14] simulated the effects of such a fleet on transit light curves for habitable planets around a variety of stars, assuming that the effective area over which starlight was intercepted was equal to the cross-sectional area of a terrestrial planet. They concluded their signature was undetectable with *Kepler*, but might be detectable with *JWST* or *TESS*. A later publication [6.20] used updated estimates of *JWST* sensitivity to conclude that detection was more difficult than previously thought, requiring ~31 transits to be observed with *JWST*.

6.2.4 Lagrange-Point Starshades

Gaidos [6.7] noted that planets orbiting sun-like stars near the inner edges of habitable zones would eventually experience runaway climate change, but could mitigate it with reflectors or scatterers near the inner

(L1) Lagrange point. Such a planetary-scale starshade would result in a characteristic maximum in the light curve at mid-transit, due to mutual occultation of the starshade and planet. Such structures would be most easily detected for rocky Earths and super-Earths around sun-like stars, but would require the capabilities of future space photometry missions to characterize transiting planets.

6.2.5 Geostationary Artifacts

Socas-Navarro [6.24] simulated the signature of a "Clarke exobelt", a very dense belt of geostationary satellites or other objects (cities, factories, etc.) in geostationary orbit around an exoplanet. He concluded that such a system might be detectable with current or upcoming technology if the surface fill fraction (i.e., the cross section of objects in an area element perpendicular to our line of sight) is of the order of 10^{-4}. This estimate was later updated with improved simulations of [6.20], who argued that a level of 10^{-3} would be necessary for a Clarke exobelt to be detectable in *JWST* observations. Figure 6.2 displays an example of a transiting Clarke exobelt surrounding an Earth analog located 1AU from a G2 star.

The left panel of Figure 6.3 shows examples of transit light curves for an Earth analog in the habitable zone (HZ) of a G2 star. It compares the planet alone, a planet with a cloud of mirror satellites with effective area equal to that of Earth, and a planet with an exobelt of sufficient density to produce a transit of similar depth. The technosignature effects will only be observable in the residuals of a "best-fit" larger isolated planet, which are shown for this example in the right-hand panel.

The geostationary orbit (GEO) is interesting because it is associated with a very specific altitude. It may be useful for a technological civilization but there are no natural processes that would place objects precisely at that particular altitude. This is important because one of the big issues in the field of SETI is how to rule out simpler natural explanations, in application of Occam's razor. The mere detection of objects in GEO would be strongly suggestive of an artificial origin.

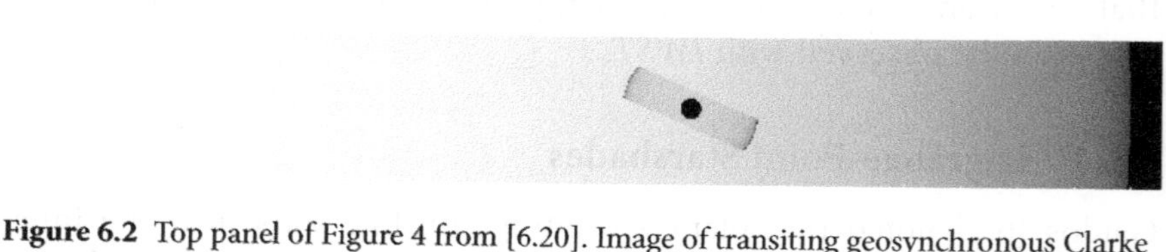

Figure 6.2 Top panel of Figure 4 from [6.20]. Image of transiting geosynchronous Clarke exobelt around an Earth-like planet orbiting 1 AU from a solar-abundance G2 star. This technosignature would be "just detectable" with ~31 JWST transits.

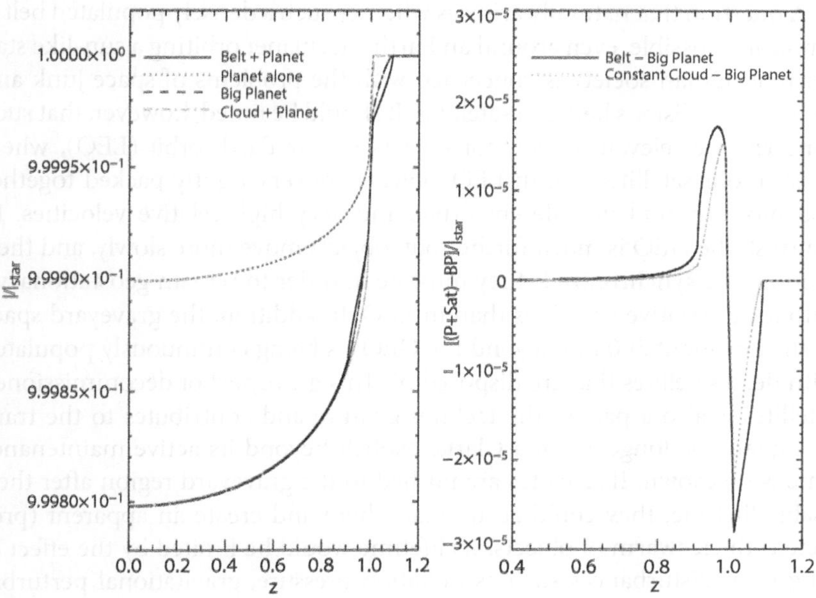

Figure 6.3 Adapted version of Figure 2 of [6.20]. Transits of an Earth analog orbiting at the inner edge of the HZ of a G2 star with half-solar abundance. *Left panel:* Transit light curves for the planet with a "just detectable" belt of size 10 R_{Earth} (solid line), the planet alone (dotted), the planet with a "just detectable" cloud as discussed by [6.14] (dash-dotted), and a bigger isolated planet producing a transit of the same depth (dashed). *Right panel:* Shows how the planet + "just detectable" cloud (dotted) and planet + "just detectable" belt (solid) light curves differ from that of an isolated big planet producing a transit of the same depth.

Given sufficient accuracy in the measurement of the transit signal, rings and moons could be distinguished from a Clarke exobelt because of the different geometry, although such imprints are probably too subtle for our current or upcoming observing capabilities. It may be possible, however, to establish that the orbiting objects are in GEO and that would imply a very high probability that we are observing an artificial system.

Synchronously orbiting Clarke exobelts are not stable if the planet itself is in synchronous rotation about its star, as the required orbit altitude lies beyond the L1 and L2 Lagrange points. Habitable planets orbiting K and M stars that host advanced civilizations are likely old enough that tidal locking will have already resulted in synchronous rotation. Thus, the existence of geosynchronous exobelts is less likely in these situations. Because of their longer tidal-locking timescales, habitable planets around G-type stars are much less likely to be in synchronous rotation and geostationary artifacts are more plausible.

A question that naturally arises is whether such a densely populated belt is physically possible, even around an Earth-like planet orbiting a sun-like star. Current human society is concerned with the problems of space junk and the risk of collisions between satellites. It should be noted, however, that such concerns are relevant mostly for objects in low Earth orbit (LEO), where most of our satellites are. In LEO, objects are very tightly packed together and move around in different orbits and very high relative velocities. In contrast, the GEO is much further out, objects move more slowly, and their motions are synchronized (they must be in order to remain geostationary), with lower relative velocities than in LEO. In addition, the graveyard space (a shell of about 300 km beyond the GEO) is being continuously populated with dead satellites that are disposed of. This graveyard of decommissioned satellites is also a part of the technosignature and contributes to the transit signal. The longevity of a Clarke exobelt beyond its active maintenance time is unknown. If satellites are pushed to the graveyard region after their usable lifetime, they could accumulate there and create an apparent (projected) thick swarm of objects. Its lifetime would be limited by the effect of long-term disturbances such as radiation pressure, gravitational perturbations (from possible moons or other planets) and collisions between objects. [6.25] considered the long-term stability of Earth's GEO orbit and concluded that gravitational perturbations from the moon, the sun and solar radiation pressure are not sufficient to deorbit or eject satellites at least on megayear timescales. The effects of collisions have been studied with simulations, showing that collision debris are expected to spread over the GEO/graveyard region (see e.g., Figures 71, 72 in [6.18]), increasing its overall opacity (the aggregate of its parts has a larger cross-section than the satellite itself).

Socas-Navarro [6.24] extrapolated the exponential growth of Earth's GEO population and found that, if this trend were to continue, our Clarke belt would be detectable from nearby stars by the year 2200 (or 2250 using the updated threshold of [6.20]) with current human technology. That milestone would mark the transition of humankind to a level of the order of 1 in the technology balance inequality discussed above. Of course, this estimate should be taken only as a rough order-of-magnitude estimate given current trends. There is no guarantee that such growth will continue, although current space developments are making launches more affordable every day and some experiments are being conducted to test technologies for a future space elevator that would facilitate access to GEO. Furthermore, proposals exist to use GEO as a launchpad for easier access to the moon [6.19] and beyond, which suggests that there might be future technological motivations for expanding current GEO usage.

6.2.6 Laser Broadcasting and Planetary Cloaking

Kipping and Teachey [6.13] discussed how an advanced civilization might affect their planet's transit profile using laser emission. This would be an effective means of signaling their presence, and would be feasible with our present technology.

However, they also note that advanced civilizations might wish to conceal their existence [6.12], and explore how their planet's transit signature could be cloaked for observers on known nearby habitable exoplanets. A fleet of mirrors like those envisioned by [6.14] could potentially be utilized, but a less elaborate method requiring less tuning would employ an array of collimated laser beams. Because the lasers only need to operate during the times of transit ingress and egress, cloaking from *Kepler*-like broadband surveys located in nearby planetary systems has energy requirements much lower than those of laser-driven light sails. Technologically, we are not far from being able to do this. However, a cloak effective at all frequencies would require a large array of tunable lasers with a much larger total power. Power needs could be alleviated by cloaking only atmospheric signatures associated with biological activity.

6.2.7 Non-Keplerian Orbits

The detection of objects following motions that significantly depart from expected Keplerian motion might be an indication of engineering. Natural departures are usually very small, such as those resulting from an object outgassing or general relativity corrections. However, artificial devices can be built to take advantage of stellar wind or radiation pressure to produce specific flight paths or configurations. Kipping [6.11] suggested that a combination of Newtonian gravity and stellar pressure could be used to maintain an artifact orbiting in conjunction with a planet, having the same orbital period but an inner orbit. He introduced the term "quasites", as an analogy to the geostationary "statites" envisioned by [6.6], to refer to some of these objects. A possible application of such systems could be for climate control, similar to the Lagrange-point starshade discussed above, although not necessarily restricted to that particular location.

6.3 Axis of Merit Discussion for These Technosignatures

The 2018 NASA Technosignature report [6.27] features axes of merit (see also [6.23]) that may be useful when designing a technosignatures search strategy. In this formulation, a search for technosignatures should be *feasible*

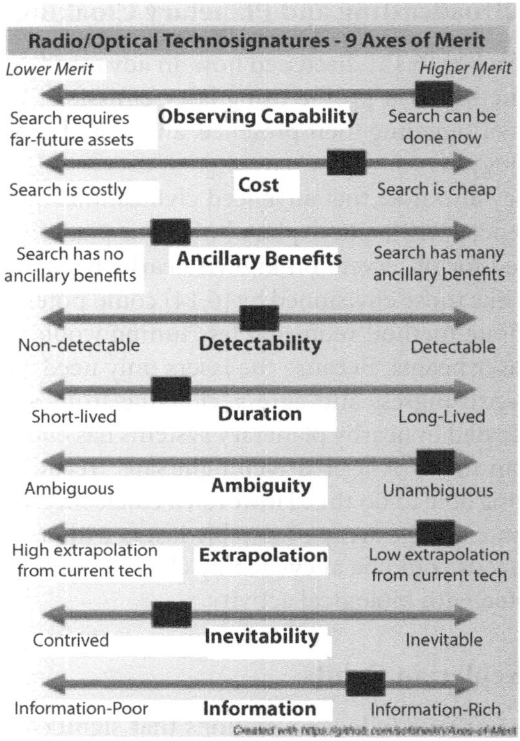

Figure 6.4 Example illustrating the axes of merit representation for optical or radio signals.

with currently available technologies, *inexpensive* to conduct and *provide ancillary benefits*. The technosignatures best suited for a coordinated search would be *long-lived, unambiguous, inevitable* and *detectable*. In addition, they would only require *low extrapolation* from current Earth technology and are preferably *information rich* when detected. Figure 6.4 shows one way of illustrating the axes of merit.

The search for technosignatures in time-series photometry scores well in many of these metrics: it can make use of currently available photometry data (from e.g., *TESS* and *Kepler*) without a lot of costly efforts, while the development of smart search techniques can bear some ancillary benefits (see section Methods below). Some of the signatures discussed here should be relatively long-lived and not require a lot of technological extrapolation from the current state on Earth. Further below we will discuss how ambiguous some of these signatures might be, e.g., in comparison to signatures of moons and other natural satellites.

In this section we briefly discuss how these technosignatures compare with one another on the axes of merit. Categorizing each one on an absolute axis of merit scale against all types of technosignatures would be ideal, but is beyond the scope of this chapter. For each technosignature, our discussion assumes that the artifacts are of the same scale as considered in the papers where they were first introduced. Frequently, when first introduced, proposed technosignatures are not quite or just barely detectable at that time. As a result, the year that a technosignature is first proposed biases its location on the detectability scale. In general, technosignatures suggested in earlier years have larger scales, so are more detectable.

A non-monolithic Dyson sphere would be the most detectable technosignature, as a consequence of its enormous size. It would rank very high in "observing capability", "ancilliary benefits" and "cost" because they could in principle be detected in astronomical surveys with existing instrumentation at great distances. However, it may not be so easy to distinguish from natural sources, such as a protostar enshrouded in dust and planetesimals, which would give it a low "ambiguity" ranking. "Duration" would be a strong point of this megastructure especially if, as proposed by Dyson, the original planets in the system have been dismantled for raw materials and therefore no large perturbers remain in the vicinity of the structure. On the other hand, it would rank very low in "extrapolation" and "information". "Inevitability" in this case is highly uncertain, given how far into the future we would need to extrapolate our technology and requirements.

Relative to other types of transit signatures discussed below, Jupiter-sized transit beacons [6.1] would rank high in detectability, since they could be seen in current *Kepler* data. They could easily be designed to be unambiguous and somewhat information rich, for example by encoding binary numbers into a louvered screen. They are far from inevitable, as they have no purpose other than signaling the presence of one civilization to another. The beacon technology also is a large extrapolation from our current capabilities.

Clouds of orbiting mirrors such as those proposed by [6.14] have an effective area similar to that of Earth. These are not quite detectable with current and upcoming missions. Made up of many smaller objects, they also require less extrapolation from our current level of technology. Since they serve a purpose other than communication, they are slightly more likely to occur than transit beacons designed solely for that purpose. However, they could be more easily confused with natural structures such as rings, and will contain low levels of information.

Lagrange point starshades [6.7] produce a characteristic light-curve maximum mid-transit, so would be more easily detected than orbiting mirror fleets, despite having a similar overall scale. Like mirror clouds, they serve a purpose other than communication, so are more likely to occur than transit beacons. At the single-object scale of terrestrial planets, they require a level of technological extrapolation between that for mirror clouds and Jupiter-sized transit beacons. Their signal would be more ambiguous than that for beacons, but less ambiguous than that of the orbiting fleets.

Given their utility for communication, geostationary artifacts ([6.24]; [6.20]) would rank high on the inevitability scale, at least for exoplanets where the geosynchronous orbit is stable. They also require a low level of extrapolation from current technology. Because they don't surround the entire planet, exobelts are less detectable than complete clouds of similar column density. Due to the uncertainty regarding technology used by other civilizations, the inevitability of an exobelt we could discover soon is quite uncertain. With light curve signatures similar to those of mirror clouds, exobelts could be more easily confused with natural structures such as rings, and will contain low levels of information. However, the discovery of a technosignature at the geostationary altitude would strongly suggest an artificial origin.

While planetary cloaking by definition will disguise or completely cancel a potential technosignature, laser broadcasting does rank well on most of the axes of merit. The search for these signals can be done with existing observatories and/or using existing archival data at relatively low (additional) costs and should be reasonably detectable, in particular when the signal is directed and repeated. These directed messages could potentially also be information rich in comparison to technosignatures that cannot carry information. They do, however, require the civilization to deliberately send these signals, which makes this technosignature quite contrived. The searches for broadcasting laser beacons can have the ancillary benefit to also find natural anomalies, while they should be easily distinguishable from them. Signals could be long-lived if the extraterrestrial civilization decides to continue broadcasting over millennia, and at the same time do not require much extrapolation from current Earth technology.

The non-Keplerian transiting objects rank high in terms of cost and ancillary benefits, since they would be found in the same way that we search for exoplanets. Detectability and observing capability would depend on the scale of the object. For instance, a starshade for climate control would be of planetary scale and would be relatively easy to detect without requiring new observing capabilities. Smaller artifacts would be unobservable with current technology. The duration of this technosignature is uncertain since no detailed studies of orbital stability have been performed to assess whether they could outlive their active maintenance time.

6.4 Methods

All of the technosignatures described above will affect the light curve of photometry for an exoplanetary system, with residuals remaining after subtracting the best-fit model of the exoplanets' size and orbital properties. Each of the technosignatures we have discussed has a characteristic signature that depends on the scale and shape of the artifacts affecting the light curve. However, comparing residuals against modelled effects for a variety of specific situations is time-consuming and will miss signatures that we have not yet considered.

A variety of approaches have been proposed to alleviate this situation. Wright *et al.* [6.31] identified 10 potential ways in which transiting megastructures could be distinguished from exoplanets, along with natural confounders for each signature. Sandford and Kipping [6.21] explored "shadow imaging": how to infer the 2-d silhouette of a transiting object from the light curve data alone, without assuming a specific model. Wright and Kipping [6.32] argue that "searching for the most unusual and anomalous signatures benefits not only the search for technologies, but also the entire astronomical community by uncovering new mysteries to advance our understanding of the Universe." Wheeler and Kipping [6.29] designed a "Weird detector" to search photometric datasets for periodic signals of arbitrary shape. Such non-parametric techniques will become increasingly useful given the growing availability of high-quality photometric surveys.

A manual search for technosignatures and other anomalies in large photometric datasets is very time-consuming and can be computationally intensive. Fast and scalable methods are needed to scan large surveys and find anomalous signals with the maximum precision at full recall (i.e., sensitivity, minimizing false negatives, detection of if possible all anomalies in the sample). This problem is tailor-made for a machine learning approach. Some promising early studies were performed by [6.8, 6.9] who used unsupervised density-based machine learning methods to detect outlier signals such as Boyajians Star in the *Kepler* dataset. Here the ancillary benefit of such a search is the detection of not only technosignatures but also naturally occurring astrophysical anomalies. See more details in the chapter by Walkowicz *et al.* (in this collection).

For signals that are too shallow to be detected in individual light curves the technique of folding has been used successfully. Sheets and Deming [6.22] for example were able to derive the average albedos of close-in super-Earths and neptunes from analysis of *Kepler* secondary eclipse data, by folding the light curves of many candidate systems for statistical detection.

In a similar fashion [6.10] folded the phasecurves of thousands of Kepler objects of interest to detect additional dips in the Lagrange points that can be interpreted as a statistical detection of the signature of Exo-Trojans on levels as low as a few ppm.

Other methods to detect technosignatures by means of transit spectrophotometry or during secondary eclipses are discussed in the literature (e.g., [6.15], [6.26], [6.16], [6.2]). For a more detailed description of atmospheric technosignatures we refer to the chapter by Schwieterman *et al.* in this collection.

6.4.1 Detectability of Time-Series Technosignatures

Each type of time-series technosignature has detectability challenges due to their time-series and spectral signatures, along with backgrounds due to similar stellar, circumstellar and planetary signatures.

The primary effects of a Dyson sphere or Dyson swarm are (1) eclipses, which can be irregular or regular in period and duration, and (2) excess infrared radiation due to the heat signature of the sphere components and waste heat due to the civilization's energy use. Of course, integrated over 4π sr, the emitted energy must still equal the energy generation within the star and sphere combined, therefore any reduction in optical flux is matched with an increase in flux in other bands, typically thermal-IR. There is no requirement that this flux be distributed isotropically, and therefore a civilization wanting to avoid detection could direct waste IR in a preferred direction, perpendicular to the galactic plane, for example. This would leave a lesser (though not non-existent) IR signature in other directions. When attempting to search for Dyson spheres, potential false positives include circumstellar dust, fully enveloped young stars, and stars with cometary or planetary debris on orbit. See [6.30] for a review of Dyson sphere technosignatures.

Orbital mirrors and Clarke belts present one of the most difficult challenges in detection of technosignatures, primarily because of the size scale and limited transit depth of these signatures. For a weak signature the additional opacity can be more simply fit by increasing the diameter of the planet by a small amount or altering the stellar limb darkening parameters. Sallmen *et al.* [6.20] found that even moderately dense Clarke belts could not be detected around any nearby G-stars with *JWST* without an unreasonable number of transit observations. A dense layer of expended spacecraft just outside of synchronous orbit could provide additional opacity [6.25]. Studies of the lifetime and capacity of such orbits are worthy of future effort. An additional difficulty in recognizing circumplanetary artifacts is the potential of rings and moons to provide similar signatures. Careful study of any candidates will likely be necessary to avoid false positives, except in cases where rings and moons are physically impossible due to dynamical constraints. Multi-wavelength observations may help distinguish the two cases.

Transit beacons will have been designed with detectability in mind, specifically to not appear as spherical planetary features. This would be done using shape, creating a transit whose depth does not match the duration of the transit entry and exit profiles, or by adding Fourier components that would not be found in natural transits, either through periodicity or addition of high-frequency components indicating a non-spherical edge.

Because they orbit the star synchronously with the planet, starshades can be difficult to detect. Depending upon the relative geometry with the line of sight, a starshade may just appear as a planet of lower density. In the best case geometry, a starshade might appear as a transit with lower depth near the center of the transit when the planet eclipses the starshade than near the edges where both block light. For most geometries, especially for stars near the outer habitable zone of low mass stars, the level of photometry necessary exceeds that possible from ground-based instruments. Serendipitous discovery is possible using *JWST* in a transit observed for other reasons.

Laser cloaking can be difficult to detect, for the obvious reason that it is designed to prevent detection. The best case for detectability is where the cloak emission matches the total stellar energy blocked, but is not spectrally identical to the stellar spectrum. For example a transit can be detected if narrow band lasers are used for the cloak. Observations at other wavelengths would still show the transit and stellar spectra during transit would show a peak at the laser wavelength. Fully replicating the stellar spectrum would require using equivalent energy to the intercepted blackbody and an emitting area similar to the disk area of the planet. As indicated earlier, this is a significant engineering challenge.

References

[6.1] Arnold, Luc. 2005, Ap J, 627, 534
[6.2] Beatty T.G. 2021, arXiv e-prints, arXiv:2105.09990B
[6.3] Boyajian, T.S., LaCourse, D. M., Rappaport, S. A., *et al.* 2016, MNRAS, 457, 3988
[6.4] Dyson, F. J. 1960a, Science, 131, 1667
[6.5] Dyson, F. J. 1960b, Science, 132, 250

[6.6] Forward R. L. 1993 "Statite: Spacecraft That Utilizes Light Pressure and Method of Use," US patent 5183225

[6.7] Gaidos, E. 2017, MNRAS, 469, 4455

[6.8] Giles, D., and Walkowicz, L. 2019, MNRAS, 484, 834

[6.9] Giles, D., and Walkowicz, L. 2020, arXiv e-prints, arXiv:2003.00109

[6.10] Hippke, M., and Angerhausen, D. 2015, ApJ, 811, 1

[6.11] Kipping, D. 2019, Research Notes of the AAS 3, 7

[6.12] Kipping, D.M., and Teachey, A. 2016, MNRAS, 459, 1233. https://arxiv.org/abs/1603.08928

[6.13] Kipping, D., and Teachey, A. 2020, arXiv e-prints, arXiv:2004.04230. https://arxiv.org/abs/2004.04230

[6.14] Korpela, E.J., Sallmen, S., and Leystra Greene, D. 2015, ApJ, 809, 139

[6.15] Lin, H. W., Gonzalez Abad, G., Loeb, A. 2014. ApJ, 792, L7.

[6.16] Lingam, M., Loeb, A. 2017, MNRAS, 470, L82.

[6.17] National Research Council. 2010. New Worlds, New Horizons in Astronomy and Astrophysics. Washington, DC: The National Academies Press. https://doi.org/10.17226/12951.

[6.18] Oltrogge, D.L., Alfano, C.L., Cacioni, A., Kelso, T.S. 2018, Acta Astronautica 147, 316

[6.19] Penoyre, Z., and Sandford, E., arXiv e-prints arXiv: 1908.09339

[6.20] Sallmen, S., Korpela, E.J., and Crawford-Taylor, K. 2019, AJ, 158, 258

[6.21] Sandford, E., and Kipping, D. 2019, AJ, 157, 42 https://arxiv.org/abs/1812.01618

[6.22] Sheets, H. A., and Deming, D. 2017, AJ, 154, 160

[6.23] Sheikh, S.Z. 2019, Nine Axes of Merit for technosignature searches, International Journal of Astrobiology, 19(3), 237-243. doi:10.1017/S1473550419000284, arXiv e-prints arXiv:1908.02683

[6.24] Socas-Navarro, H. 2018, Astrophys.J. 855, 110

[6.25] Socas-Navarro, H. 2020, Technoclimes Workshop, https://www.youtube.com/watch?v=saT8w0YBTT0

[6.26] Stevens, A., Forgan, D., James, 2016., International Journal of Astrobiology 15, 333.

[6.27] Technosignatures Workshop Participants 2018, NASA and the Search for Technosignatures: A Report from the NASA Technosignatures Workshop, arXiv e-prints arXiv:1812.08681

[6.28] Verendel, V. and Häggström, O. 2017, International Journal of Astrobiology 16, 14

[6.29] Wheeler, A., and Kipping, D. 2019, MNRAS, 485, 5498. https://arxiv.org/abs/1903.09213

[6.30] Wright, J. T. 2020, Serbian Astronomical Journal, 200, 1-18 https://arxiv.org/abs/2006.16734

[6.31] Wright, J. T., Cartier, K. M. S., Zhao, M., Jontof-Hutter, D., and Ford, E. B. 2016, ApJ 816, 17
[6.32] Wright, J., and Kipping, D. 2019, BAAS, 51, 343 https://arxiv.org/ftp/arxiv/papers/1907/1907.07830.pdf

7

Post-Detection Message Analysis and Comprehension

Brian S. McConnell

SETI Open Data Archive, San Francisco, CA, USA

Abstract
The chapter explores the message analysis and comprehension efforts that would follow a SETI detection event and discusses the information types and patterns we might look for in an information-bearing signal from an extraterrestrial intelligence.

Keywords: SETI, METI, extraterrestrial intelligence, first contact, information theory, communication systems

7.1 Categorizing an ET Signal

The first step in comprehending an ET signal will be to understand the type and intent of the communication. We will want to know if the signal modulated to transit information, and if so, is the information intended for us?

7.1.1 Megastructures

Technosignature surveys are searching for evidence of megastructures such as Dyson Spheres. A swarm of orbiting solar arrays will be detectable to astronomers, as the array would block some of the visible light emitted by the star, and re-emit light in the infrared spectrum, resulting in an excess of infrared in the spectrum for the star. This type of signal would tell us that another civilization exists or had existed at that location.

Email: bsmcconnell@gmail.com

Anamaria Berea (ed.) Technosignatures for Detecting Intelligent Life in Our Universe: A Research Companion, (103–126) © 2022 Scrivener Publishing LLC

7.1.2 Beamed Power and Propulsion Signals

Technosignature surveys also look for evidence of high-power transmissions that cannot be explained via natural processes. These may include beamed power or propulsion systems, similar to the Breakthrough Starshot system [7.1], which will use an array of high-power lasers to propel small lightsails to 0.2c. Systems like these would appear to us as high-power, monochromatic optical or microwave signals, unlike what would be expected from a natural source. This type of signal would not necessarily be designed to transmit information to the target, so were we to detect one aimed at us, we would learn that a technological civilization existed, but may not be able to extract information from the signal unless the propulsion beam was also modulated to transmit information.

7.1.3 Intercepted Communication

Technosignature surveys may also detect an information-bearing signal that is intended for someone else. This type of signal may carry a large amount of information, but be encoded in a way that precludes naive parties from extracting data from the signal.

Very long-range communication requires the use of forward error correction to ensure the reliable delivery and recovery of data at the receiving end without requiring the receiver to request retransmission. Forward error correction works by employing algorithms to encode the transmission such that even if part of the transmission is corrupted or lost, the original data can be recovered without retransmission.

The problem this poses for anyone wishing to eavesdrop on the communication is that without prior knowledge of the algorithm used to encode the transmission, it will be difficult or impossible to reconstruct the transmitted information.

7.1.4 Intentional Communication

This is the classic SETI contact scenario, where a civilization is transmitting to us with the intention of sharing information or initiating communication. This type of signal will differ from others as it will likely be designed to facilitate the process of parsing and comprehending the signal and the information conveyed.

While it is impossible to predict what specific information an ETI might include, we can anticipate the technical challenges associated with interstellar communication, as well as the techniques likely to be used to counter them. The designer of an interstellar communication link will need to deal with the following challenges:

1. Latency due to the speed of light. In the most optimistic scenarios involving the nearest solar systems, round trip communication between systems is on the order of a decade. This is quite unlike terrestrial communication where maximum latency is on the order of a fraction of a second for the longest links (e.g., geostationary satellite relays).
2. Transmission errors and drop outs. The transmission link should be considered unreliable, with errors and drop outs due to background interference, occlusion, and planetary rotation and other causes.
3. Duty cycle mismatch. The receiver is very unlikely to detect the transmission just as it begins and receive it in its entirety. The sender should assume that the receiver will only receive part of the transmission during each pass.
4. Naive receiver. The receiver has not received communication from the sender before, and thus is unaware of the methods used to compose and organize the information transmitted.

To deal with these challenges, the sender will need to design the communication channel such that the receiver can do the following:

1. Reconstruct corrupted or missing data without requesting retransmission. This is known as forward error correction and is widely used in communication systems built upon unreliable communication channels.
2. Reassemble a large message or collection of elements from smaller sub-elements, which need not be received in sequential order.
3. Understand at least some parts of the transmission even if other parts of the transmission remain obscure or difficult to comprehend.

These design criteria favor the use of something akin to a digital packet switched network, where a large block of information is broken down into parcels that are labeled such that they can be reassembled and so that errors can be detected and corrected on the receiving end.

7.2 The Interstellar Communication Relay

Following a detection event, especially in the case of an information-bearing signal, SETI organizations will need to build and operate a system for archiving and distributing the information extracted from the signal.

This system—we'll call it the Interstellar Communication Relay—will enable a large number of people and organizations to access copies of the information received. Because the transmission rate will likely be low during early stages of this process, the technical and financial resources needed to operate this system will be modest. Cloud-based storage and computing platforms, such as Google Compute and Amazon Web Services, will enable SETI organizations to build highly scalable systems that can accommodate millions of concurrent users at a modest cost.

7.3 The Processing Pipeline and Participants

Unlike the initial detection effort, which will be dominated by astronomers and signal processing experts, the message analysis and comprehension effort will involve people from a broad field of studies, and will include people outside of academia. It is possible the best insights will come from people who are not presently affiliated with SETI.

The ICR will enable the creation of a data processing pipeline that involves the efforts of subject matter experts from many fields, as shown in Figure 7.1.

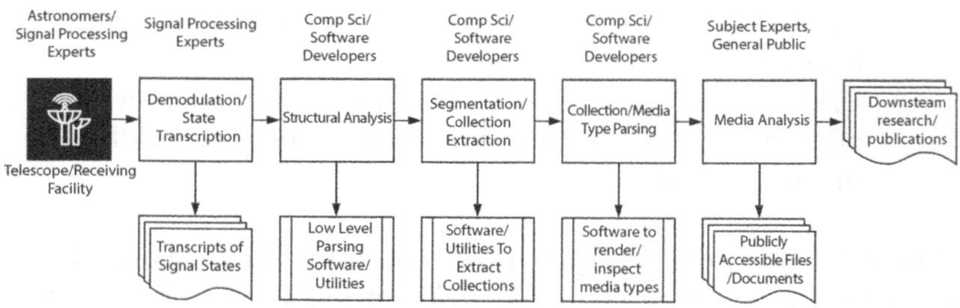

Figure 7.1 Flowchart of the message analysis pipeline, with participants and outputs for each step.

7.4 Demodulation

Researchers will investigate whether the signal is modulated to transmit information. For the purposes of this chapter, we will consider a pulsed laser signal, as the possible modulation techniques can be described in less technical terms.

Optical SETI surveys look for brief bursts of photons, on the order of a nanosecond. A laser that concentrates its power into a short burst can briefly outshine its background star, and thus be detectable at interstellar distances.

This type of pulsed beacon can be modulated in several ways to transmit information.

7.4.1 Pulse Width Modulation

The duration of each pulse can be used to represent information. Consider an example where each pulse has one of four durations and represents 2 bits of information. The receiver will notice that the pulse durations cluster around these values and can infer the number of possible signal states per pulse.

7.4.2 Pulse Interval Modulation

A pulsed laser beacon will not be transmitting for the vast majority of time, so additional information can be transmitted by varying the interval

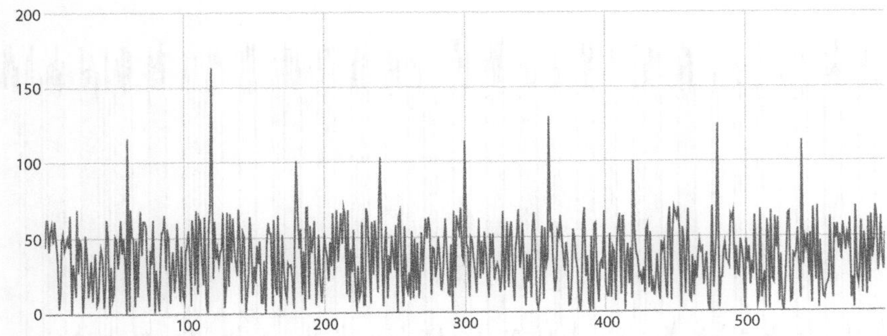

Figure 7.2 Sample plot of photons counted over time. Notice the spikes that rise above the background photon counts. This is the type of pattern expected in an Optical SETI detection event.

between pulses. Consider an example where individual pulses are spaced between 50 milliseconds and 150 milliseconds, with 5 nanosecond increments in the possible pulse intervals. This results in 20,000,000 possible values for a pulse interval. The transmitter could encode up to 24 bits of information in this case (Figure 7.2).

7.5 Combining Pulse Width and Pulse Interval Modulation

7.5.1 Wavelength Multiplexing

The sender can further increase the information carrying capacity of the system by using several lasers, each tuned to a different wavelength, each with its own pulse width and pulse interval modulation subchannels. Let's assume the sender is transmitting pulses on four different wavelengths in this case.

7.5.2 Information Carrying Capacity

Let's estimate the information carrying capacity of this hypothetical system (Figure 7.3). There are four color channels, and two subchannels (pulse duration modulation and pulse interval modulation) for each. Each laser pulses on average ten times per second. The information carrying capacity can then be estimated as:

$$I = (intervalbits \times n_{colors} \times avgpulserate) + (widthbits \times n_{colors} \times avgpulserate)$$

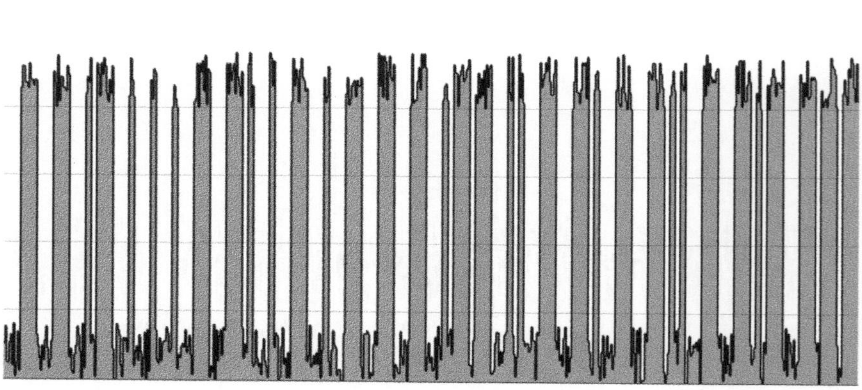

Figure 7.3 A plot of a signal that combines both pulse width and pulse interval modulation, with two possible values for each subchannel (1 bit/subchannel).

This example results in an information carrying capacity of 1,040 bits per second. While this may seem slow compared to terrestrial internet connections, it is possible to transmit roughly 32 gigabits of information per year, the equivalent of several thousand high resolution photos.

7.6 Data Extraction

7.6.1 Transcription

As teams begin extracting data, this information will be transcribed into a format that is suitable for further analysis. They will not know what the information represents except that each signal state can represent N number of symbols. The signal may be organized as parallel communication channels, such as the case of a pulsed laser beacon using wavelength multiplexing.

The data product from this stage of the data processing pipeline will be a set of files that assign arbitrary characters to different signal states. A hexadecimal text file with interleaved metadata is a good example of an intermediate file format, as shown below.

```
<facility>OSETI.1</facility>
<channel>600nm</channel>
<subchannel>pwm</subchannel>
<timestamp>2020-05-19 12:52:19.880523</timestamp>
<data base=16>78e919ad23a120c3b074922a11fac-
c3b551c951fee3ff5f1809a257506916315dead-
7ee5a32593ee08ee7203c82c42aa-
1511faa5387a7e812eb4cbfce9e4f584</
data>

<facility>OSETI.1</facility>
<channel>600nm</channel>
<subchannel>pwm</subchannel>
<timestamp>2020-05-19 12:52:20.887200</timestamp>
<data base=16>2e43d27c130495dfc20de50142ef
5982e61591970b86b52b1d3a3ac37a49931660e-
b93eb76064d3cca2d8d79cda84ff239ec-
37cd6d6a161ef7ba335d8e25fd05</
data>
```

7.6.2 Structural Analysis

The next step will be to analyze this data to look for repeating patterns or structure. If, for example, the transmission is segmented into smaller sub-elements, researchers will want to know what states represent the equivalent of START or END symbols or something akin to parentheses, as these can be used to break a large block of data into smaller sub-elements or collections.

7.6.3 Segmentation and Extraction

Once the low-level encoding scheme(s) are known, the next step in the processing pipeline will parse the data stream and split it out into different collections and segments (this assumes the transmission is a multipart data set). This is analogous to how a computer file system works as it reads serially recorded data from a storage medium and represents this to users as a collection of files and collections. The goal at this stage is not to understand what any particular collection or message element represents, just to recognize the boundaries between elements and extract them as separate collections of data.

7.6.4 Data Types

Once analysts have succeeded in breaking the data stream into smaller elements, others can then begin the work of understanding what different elements represent. Digital communication systems are useful because they can represent anything that can be represented in digital form, including images, time domain data such as audio, executable programs, etc. A multipart message will need to provide a mechanism for splitting segments out as discussed before, as well as for differentiating between different data types. What media types might we encounter in a multipart message?

It is worth noting that in discussing this, we are anticipating what modes of communication an astronomically literate civilization may be capable of, not necessarily what modes of communication another species prefers. SETI and interstellar communication are branches of astronomy, and while their goal is to detect technological civilizations, there is an implicit assumption that they are looking for other astronomically literate civilizations. Knowledge of astronomy is a prerequisite for interstellar communication. An intelligent species might prefer different sensory modalities to vision for their own use, but in order to be proficient in astronomy they would need to understand light and photography.

The sending species might sense its environment via echolocation, much like bats here on Earth. While they might prefer this sensory modality, they would need to understand electromagnetic radiation and photography, and to build sensory apparati (e.g., telescopes), in order to be astronomically literate. This is not unlike how humans have created ultrasonic imaging technology to see via echolocation.

7.6.5 Communicating Observables vs. Qualia

Communicating observables such as images or audio samples should be straightforward since these can be represented numerically. Describing qualia or internal experiences may prove impossible due to the lack of common sensory modalities. The senses of smell and taste are good examples. Both are intimately linked to the viewer's physiology and internal experiences. How would you describe the concept of sweetness to someone who can't experience it?

With this in mind, we can anticipate some of the modes of communication that might be employed in an interstellar communication channel, and that will allow some degree of communication between parties with differing capabilities and native modes of communication.

7.6.6 Images

The ability to capture and process images is a prerequisite for astronomy. Uncompressed mages are trivially easy to represent in a digitized format as an array of numbers, and should also be relatively easy to decode as the number of parameters used in their encoding are relatively small. These include the height/width dimensions of the image, number of bits used per pixel, number of color channels, planar vs. polar projection, etc.

The sender can use images of common or mutually observable astronomical objects to assist the receiver in understanding how to decode and render images. Planetary images (Figure 7.4) will be useful as this type of image consists of a spherical object against an empty or black background. This pattern is easy to recognize, even if the image data is embedded in a larger bit stream.

The sender can take this a step further by including images of astronomical objects such as nebulae and supernovae remnants that are mutually observable at great distances. These images can be used to finely calibrate how images are decoded and rendered on their end (Figure 7.5).

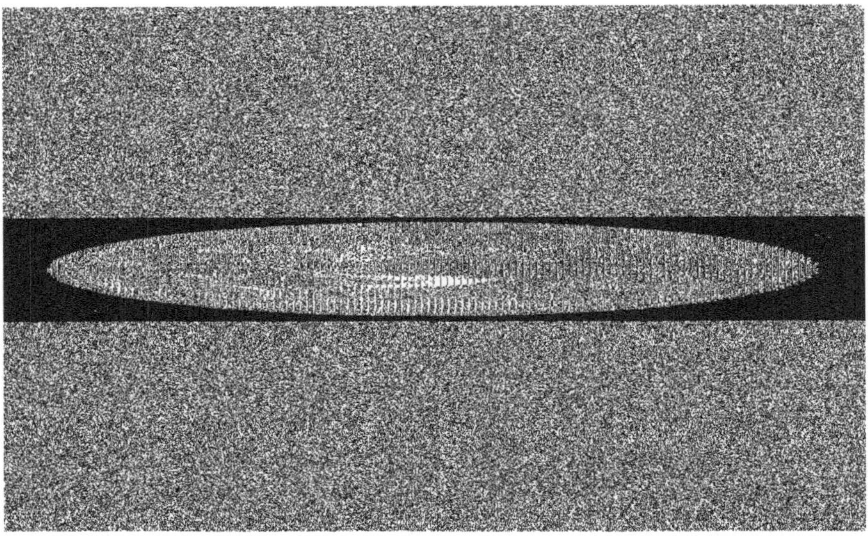

Figure 7.4 Example of a planetary image embedded in a larger block of random data. Notice how the empty regions around the planetary object stand out. The receiver only needs to guess the width of the image for this pattern to be clearly visible, and can proceed to work out the rest of the encoding parameters from there.

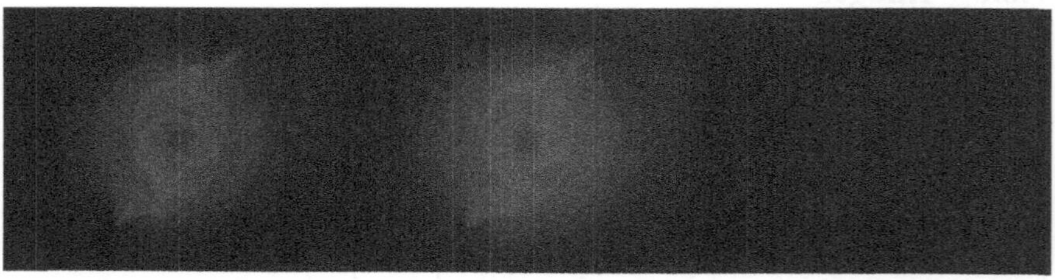

Figure 7.5 The Cat's Eye nebula, imaged in red, green and blue color channels. The receiver can compare these images against their own observations at various wavelengths to find the closest fit for rendering images. Image credit: NASA, ESA, Hubble Heritage Team [7.2].

7.6.7 Time Domain Signals

Time domain signals, when represented digitally, consist of a series of numeric samples. This format is commonly used to record sound and other data series. A typical audio sample will consist of a series of N-bit samples, where the sample size determines the sample resolution. The rate at which samples are taken determines the frequency range of the audio stream, where the maximum frequency is roughly one-half the sample rate.

Consider a raw audio stream, with no additional processing such as companding, which is used to increase dynamic range. To decode this type of audio sample, the receiver only needs to guess the sample size and sample rate. The sender can assist the receiver in decoding audio by using easily recognizable reference cases.

A sine wave (Figure 7.6) will enable the receiver to work out the sample size used for encoding, because incorrect guesses at sample size will play out as static or noise while the correct guess will play out as a pure tone. This parameter space is small so the receiver should be able to figure out the sample size used for this type of reference signal. The sample rate is more difficult to guess because it is measured in terms of time. So how might the sender give the receiver a clue about the time base they are using?

Here again mutually observable astronomical objects such as pulsars can be put to use. An astronomically literate civilization that is capable of interstellar communication will have likely developed the ability to detect and study them. As a point of interest, the first pulsar humans detected, PSR B1919-21, was discovered in 1967 by Martin Hewish and Jocelyn Bell Burnell. The signal was so regular that one explanation scientists considered was that it was an extraterrestrial beacon.

By including samples of pulsar signals (Figure 7.7), the sender can assist the receiver in understanding the time base used for digital audio streams. The receiver could match the sampled signal against their observations of

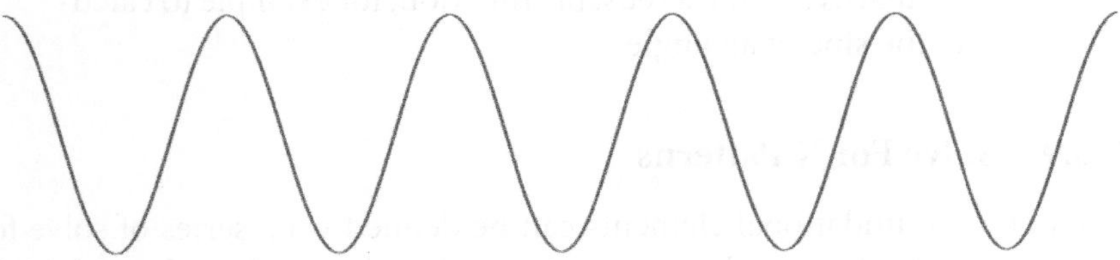

Figure 7.6 A pure tone (sine wave) as would appear in a time domain graph (x-axis → time, y-axis → amplitude).

Figure 7.7 A 16-bit binary representation of the signal plotted above, in unsigned integer notation. Notice the repeating pattern in the bit field from left to right.

the same pulsar to precisely determine the sample rate. The receiver would then be able to decode any audio stream using that time base.

The sender can also help the receiver by including audio samples of phenomena that are common to worlds with liquid oceans such as breaking waves, thunder, rainfall, etc. While these sounds will vary due to atmospheric pressure and composition, it will be possible for the receiver to narrow the time base down to a range of likely values.

It is worth noting again that the inclusion of audio data need not imply that the sender has a natural sense of hearing, only that they understand how to record audio as a sequence of pressure readings, and might reasonably expect their counterparts to have a similar scientific understanding of it.

7.6.8 Algorithms (Computer Programs)

Algorithms are another data type we may encounter in an ET communication channel. They can be built using a small collection of math and logic symbols which can be taught via "solve for x" examples.

To define a programming language, one needs the following elements:

- Basic math operators: addition, subtraction, multiplication, division, modulo/remainder
- Comparison and conditional branching: if a > b, do x; else do y
- Read/write from memory: peek(n); poke (n, value)
- Modularity: define a reusable function, for example to calculate the sine of an angle

7.6.9 Solve For X Patterns

Each of the foundational elements can be defined via a series of solve for x statements that guide the receiver toward understanding the unknown x in each series. A human readable example of this might look like the following.

2 ? 2 = 4
3 ? 1 = 4
1 ? 3 = 4
0 ? 4 = 4
4 ? 0 = 4
? → addition

2 ? 1 = 1
3 ? 1 = 1
4 ? 1 = 1
1 ? 2 = 0
1 ? 3 = 0
? → is greater than

This general pattern can be used to build up a small lexicon of the basic math and logic operations needed to define a modular programming language.

7.6.10 Memory

An extensible programming language should also be able to read and write from persistent memory. This allows a program or function to record its outputs for future use and vice versa. A program in an ETI message might read from static data contained elsewhere in the datastream. An image decompression program, similar to a JPEG viewer, is a good example of this pattern. In this case, the sender defines the algorithm to decompress image data, which can then be used to render any number of compressed images found elsewhere in the datastream.

To do this, the programming language should include operators to read from and write to a virtual memory address. The peek and poke functions in early forms of the BASIC programming language are a good example of what this might look like. With these two functions, a program can read and write to memory outside of the program as follows:

```
peek(n) : reads data from the memory location n
poke(n, v) : writes the value v to the memory location n
```

7.6.11 Modularity and Reuse

An extensible programming language should also enable its users to build higher-level functions that can be reused and incorporated into other functions. The function to calculate the approximate sine of an angle is a good example as it can be calculated through a combination of simple arithmetic and conditional branching by using the Taylor series, as shown below:

$$sine(a) \approx \frac{a^1}{1} - \frac{a^3}{3!} + \frac{a^5}{5!} - \frac{a^7}{7!}$$

in expanded form

$$sine(a) \approx \frac{a}{1} - \frac{a \times a \times a}{3 \times 2 \times 1} + \frac{a \times a \times a \times a \times a}{5 \times 4 \times 3 \times 2 \times 1} - \frac{a \times a \times a \times a \times a \times a \times a}{7 \times 6 \times 5 \times 4 \times 3 \times 2 \times 1}$$

This sequence of instructions can be expressed in the Python programming language using the code below (note that we assume the functions power() and factorial() have been defined previously).

```
def sine(a, steps = 9):
  a = a % (2 * 3.14159265359)
  v = 0
  n = 1
  p = 1
  while n < (steps * 2):
    if p == 1:
      v  = v + (power(a, n) / factorial(n))
      p  = -1
    else:
      v  = v - (power(a, n) / factorial(n))
      p  = 1
    n = n + 2
  return v
```

The function above calculates the approximation for the sine of the angle a, where a is given in radians, and steps is the number of steps in the Taylor series (if omitted, it defaults to 9 steps). Once defined, the user can simply use the expression below whenever they need the sine for an angle.

```
y = sine(a)
```

Modularity eliminates the need to repeat the instructions for a function every time it is used. This reduces the size of the programs, which is vital given the constraints in interstellar communication. This also enables users to create higher-level functions using reusable functions as building blocks, all of which reduce to the small lexicon of math and logic operations used to define the language.

It is also important to note that there are many ways to construct a Turing complete language. The Python example above is shown for readability. An example of a fully worked out algorithmic communication system for SETI is Paul Fitzpatrick's CosmicOS [7.3]. It is based on lambda calculus, and is not as easily read, but can do the same things a Python-like language can.

7.6.12 Symbolic Communication Systems and Artificial Languages

While communication via natural language between species is difficult, it may be possible to communicate higher-level concepts via an artificial or constructed language. Semantic networks are an example of how this type of communication might work.

A semantic network is a graph of how ideas or concepts relate to each other. Each unique concept is assigned a unique numeric address, not unlike the IP address of a computer on the internet. The address itself has no meaning. It is just a number that uniquely identifies a concept so it is not confused with others. Meaning is expressed by mapping how each concept is related to others using operators such as: is a member of, is equal to, is similar to, is not similar to, is greater than, etc. Semantic addresses can also be linked to other representations such as images or audio to link them to representations of real-world objects.

An important feature of this type of system is disambiguation, whereas a word in a natural language can have many different meanings depending on its usage and context. In a numeric semantic network, each meaning or use has a different address. The address space is given by 2^n for a

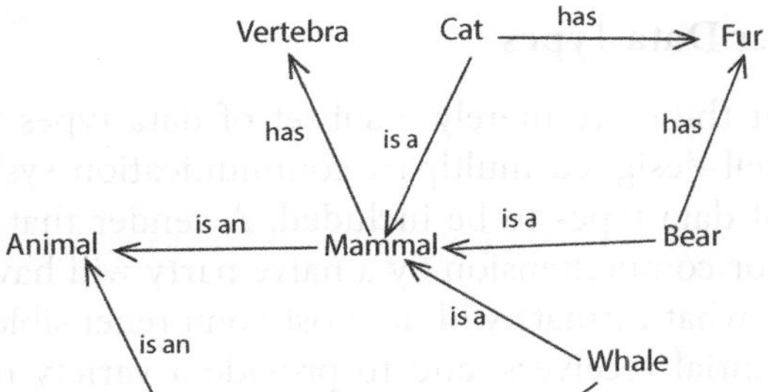

Figure 7.8 An example of an English semantic network. Notice how each concept is linked to others via operators such as "is a(n)", "has", etc. A numeric semantic network functions in the same way except concepts are labeled with unique numeric identifiers. Image courtesy: Wikipedia.

binary system. A 32-bit word size would allow for approximately 4 billion unique semantic addresses. Slightly larger word sizes, such as 48 or 64 bits, result in very large address spaces of $2.8x10^{14}$ and $1.8x10^{19}$ respectively. Accommodating a large vocabulary of concepts will not be a problem in such a system.

A semantic network (Figure 7.8) can be used as the foundation for a constructed language in which expressions are created by using semantic addresses in place of words. Because the system is numeric in nature, it will not be dependent on either party having shared modes of natural communication. This constructed language can be thought of as a machine readable interlingua that can be transformed to and from natural language equivalents on either end of the communication link.

7.6.13 Genomic Information

DNA sequences are composed of a series of four base pairs (2 bits per symbol) and as such serve as a digital storage medium. A sender might choose to include genomic information for a variety of reasons, for example to assist the receiver in understanding the evolutionary history of organisms in their world.

Transcribing this information to a digital format is straightforward, though recognizing this information for what it is may be difficult. We should not assume that DNA is a universal information carrier, but rather that other life systems will likely need a similar system to store and transmit genetic information, and that it is possible a digital representation of this information may be included.

7.6.14 Other Data Types

It is likely that these are merely a subset of data types which may be included. A well-designed multipart communication system will allow for a variety of data types to be included. A sender that is designing a transmission for comprehension by a naive party will have an incentive to think about what formats will be most comprehensible to the widest variety of potential receivers, and to provide a variety of media types and paths to comprehension to work around differences in sensory modalities.

7.7 Classifying ETI Communication

The elements of an ET communication system can be classified by the level of interactivity with its receiver.

7.7.1 Static/One Way Communication

A static message, like a printed page, is non-interactive and relies on the receiver to parse and comprehend its contents. Examples include pixelated images, linear signals such as audio samples, semantic networks and texts. These may be straightforward to extract and are good starting points for communication, but they are limited to one-way communication and are

Figure 7.9 A river flowing with mountains in the background. Image credit: "The Tetons - Snake River", Ansel Adams. This image was included in the collection on the Voyager record [7.4].

Figure 7.10 Example of a labeled image from the Voyager record which depicts the Golden Gate Bridge, annotated to indicate its scale. Image credit: "Golden Gate Bridge", Ansel Adams Publishing Rights Trust, annotated by Jon Lomberg *et al.* for the Voyager record [7.5].

limited in what they can describe on their own. Consider Figure 7.9, which depicts a river in a mountainous area.

Figure 7.10 may be difficult to interpret in isolation. There is no information about the scale or context of the objects in the photo. It is obvious to a human viewer because of our shared experience with this type of terrain. A viewer from a mostly oceanic world with no mountains might misinterpret the objects within the scene as well as their scale.

The sender can assist the receiver by annotating images, for example by adding labels to indicate relative sizes, associate objects within a scene with concepts in a semantic network, etc. The designers of the Golden Records flown on the Voyager space probes developed a system of annotations that were overlaid onto images to highlight key concepts and relative scale.

Audio samples pose similar challenges. Some sounds will be familiar to most listeners. The sound of breaking waves should be familiar to anyone from an oceanic world. Other sounds, such as bird songs, may be unique and thus unfamiliar.

The primary limitation of static data types is that they are non-interactive, so the receiver will have no way of probing them for more information or clarification.

7.7.2 Algorithmic Systems

A message composed of algorithms will have a number of important features compared to static, non-interactive communication. Among them:

1. Programs and functions can be defined using a small set of basic math and logic instructions. These functions can be executed on a virtual machine or interpreter, without the need to define a specific computing architecture or substrate.
2. Progressively more sophisticated and reusable instructions can be built on this foundation.
3. Programs within the message can implement arbitrarily complex compression and error correction algorithms, and thus provide the foundation for a very robust communication channel that operates near the theoretical limits of efficiency.
4. These programs will be able to interact with their users in real time, and therefore avoid the constraints of speed of light delays in interstellar communication.

The capabilities of these programs will depend on the sophistication of their authors. While these programs would reduce to a small set of basic math and logic instructions, they may exhibit complex behavior.

Algorithms can be further classified by their complexity and ability to interact with and learn from their users.

7.7.3 Statically Defined Programs

These are stand-alone programs designed for a specific purpose. They do not retain information and merely execute a predetermined set of rules. These programs, while simple, can model a very large number of possible states. A program that implements the game GO is a good example. The rules for the game are simple and can be expressed in a compact program, yet the game itself has a near infinite number of possible states to explore. These types of programs will be useful in explaining physical systems and scenarios that can be modeled in a game-like environment.

7.7.4 Data-Driven Programs

This class of program interacts with data outside of the program's instruction set. In terrestrial applications, programs often interact with an external database. A program might also record data related to its usage which may affect the behavior of the program in future sessions, a rudimentary form of learning.

In the context of ET communication, the sender might construct programs that assist the user in parsing data elsewhere in the transmission. The sender might include image processing programs that implement robust image compression/decompression algorithms, and thus greatly expand the number of images that can be transmitted via a bandwidth limited communication channel.

7.7.5 Machine Learning/Narrow AI

While our experience with computation is recent, we have already begun building precursors to artificially intelligent systems. These machine learning programs are modeled after nervous systems. Instead of being programmed with logical instructions, they are trained using data sets, and learn to recognize patterns as they are exposed to examples. These programs can already match or exceed human capabilities at specific tasks such as image classification and speech recognition. These narrow AIs are best thought of as specialists that are optimized to perform a particular task, such as recognizing animals in photographs. They are not intelligent in the sense that a person is, but can exhibit superhuman capabilities in their area of specialization.

7.7.6 Adaptive/Self-Learning Systems

The most sophisticated programs we might encounter in an ET transmission would be adaptive or self-learning programs that can be thought of as a form of artificial intelligence. We do not yet know whether it is possible to build a general AI, an intelligent system that rivals the flexibility or repertoire of human thought. An extraterrestrial civilization is likely to be quite advanced relative to human civilization, and will have had much time to advance its science of computation. We should consider the possibility that an ET transmission may contain elements that are themselves intelligent or possibly even sentient.

The presence of such an agent within a transmission would pose another set of issues and risks compared to simpler forms of communication since it would be difficult or impossible to anticipate the behavior or motivation of the agent.

7.8 Risks of Contact

While there has been much discussion about the risks posed by outbound communication or Messaging Extraterrestrial Intelligence (METI), there has been less discussion about the risks posed by receiving information. These risks are not insignificant, both in terms of risks posed by human behavior in response to contact and risks posed by the communication itself.

7.8.1 Internal (Human Originated) Risks

7.8.1.1 Interstate Competition and Interference

State actors may seek to monopolize or deny access to an ET signal, for example by disrupting or destroying a competing state's receiving facilities. This scenario is plausible especially if a state actor sees military or economic value in obtaining exclusive access to this information [6].

7.8.2 Misinformation

Another risk posed by the detection of an ET transmission is that bad actors such as cults and con artists will attempt to manipulate the flow of information or claim special access to the communication. The internet and social media have enabled the spread of false or misleading information on a global scale, and we should expect that this will occur in the wake of a detection event. Of particular concern is a scenario where analysts are unable to parse the information received, as this will not deter third parties from manufacturing false narratives in the absence of scientifically confirmed communication.

7.8.3 Societal Reaction

How society at large will respond to contact is unknown, and is a subject that has been explored in depth via science fiction (Arthur C. Clarke's *Childhood's End* is a personal favorite). The nature of the response will probably depend upon the type of information communicated, and also how it is perceived by humans. Even if the intent is benign, we may misinterpret it as threatening.

7.8.4 External Risks

The external risks posed by receiving information from an extraterrestrial intelligence should also be taken into consideration. Here the worry is less about physical contact resulting from one-way communication than from dangers posed by the information itself.

Static information, such as pixelated images, poses little direct risk since it is not autonomous in any sense. The primary risks stem from algorithms and genomic information contained in a transmission, if present.

7.8.5 Algorithms

7.8.5.1 Malware

A risk posed by ET algorithms is that they may be designed to behave in a manner similar to a computer virus or worm. This seems unlikely since it would run counter to the purpose of establishing communication with the receiver, but we should take precautions when examining algorithmic elements of an ET communication.

The best practice will be to run them on simulated computers that have no network connectivity. This will mitigate the risk that an errant program will be able to escape the confines of its simulated environment. Each element would be examined in isolation and once its behavior is understood could be approved for wider use.

7.8.6 Artificial Intelligence

General AIs, if present, could pose significant risks. The fundamental challenge posed by AIs is that their internal workings are not easily inspected and understood without prior knowledge of how the systems are built.

We can only speculate about what the motivations of a general AI would be. Its motivation could be to help the receiver understand what is being communicated. It could be self-interested and primarily concerned with reproducing itself. It could be malevolent. We would have no way of knowing. Because of such unknowns, such a program or system would need to be isolated on a virtual machine so that it could not directly manipulate anything outside of that environment.

7.8.7 Genetic Information

Genetic information, if present, should be approached with caution. We do not know yet how life functions elsewhere. Perhaps DNA has emerged independently on different worlds as an information carrier, or maybe each world has developed different systems for transmitting genetic information. This alone is valuable information, and is something we could potentially learn from ET communication.

Genetic information will be valuable in classifying life and understanding the evolutionary history of a world. Imagine a catalog of a world's organisms. The sender could include pictures, but pictures only reveal an organism's visible elements. By including the organism's genome or parts thereof, the sender can reveal a great deal more, such as what parts of its genome are shared with other organisms, and thus its evolutionary history.

There is little risk in studying this information digitally; however, there would be significant risk in using this information *in vivo*, especially in the case of RNA or DNA sequences, if present. It goes without saying that it would be a bad idea to transplant this information into any terrestrial organism.

7.8.8 Unknown Unknowns

Beyond obvious risks, such as those posed by algorithmic systems, there are likely unforeseen risks that we have not yet taken into account, especially in terms of misinterpreting the nature of the communication or projecting human intentions onto it [7.7].

Unfortunately, it is difficult to predict societal effects as historical examples are imperfect analogies that involve contact between different human civilizations that, despite their cultural and technological differences, had much in common.

References and Recommended Reading

[7.1] Breakthrough Starshot website, viewed May 26,2020, https://breakthroughinitiatives.org/initiative/3

[7.2] Cat's Eye Nebula, courtesy NASA, ESA, Hubble Heritage Team, June 10, 2018, https://apod.nasa.gov/apod/ap180610.html

[7.3] Fitzpatrick, Paul, "Cosmic OS", MIT CSAIL website, https://people.csail.mit.edu/paulfitz/cosmicos.shtml

[7.4] Adams, Ansel. "The Tetons - Snake River, Grand Teton National Park" (1942), Reproduced with permission from The Ansel Adams Publishing Rights Trust. All rights reserved. https://catalog.archives.gov/id/519904

[7.5] Adams, Ansel. "Golden Gate Bridge" (1953), Reproduced with permission from The Ansel Adams Publishing Rights Trust. All rights reserved. Annotated by Jom Lomberg et al for the Voyager records.

[7.6] Wisian, Kenneth W. and Traphagan, John W. "The Search for Extraterrestrial Intelligence : A Realpolitik Consideration". (2020) Space Policy, https://doi.org/10.1016/j.spacepol.2020.101377

[7.7] Traphagan, John W. "Anthropology at a Distance : SETI and the Production of Knowledge in the Encounter with an Extraterrestrial Other" (2014) In: "Archeology, Anthropology, and Interstellar Communication." Vakoch, Douglas A (ed.) The NASA History Series, Washington DC. 131-142.

8

Statistical Issues in the Search for Technosignatures

Amedeo Balbi[1]* and Claudio Grimaldi[2]†

[1]Dipartimento di Fisica, Università di Roma "Tor Vergata", Via della Ricerca Scientifica, Roma, Italy

[2]Laboratory of Physics of Complex Matter, Ecole Polytechnique Fédérale de Lausanne, Station 3, Lausanne, Switzerland

Abstract

New ambitious initiatives searching for technosignatures from exo-civilizations are currently being pursued and others are planned in the future, allowing exploration of the parameter search space on an unprecedented scale. Despite this intense activity and the decades-old history of related searches, there have been only limited efforts in analyzing systematically the prospects of either the detection or the non-detection of technosignatures. This chapter will review the recent progress made in developing full physical-statistical descriptions of electromagnetic technosignatures that could possibly populate the Milky Way.

Keywords: Technosignatures, SETI, drake's equation, theory of probability, bayes' theorem

8.1 Introduction

While life appears to have emerged relatively quickly on Earth, humans and their technology arose only billions of years later, a fact sometimes taken as evidence for a low emergence rate of technological species beyond Earth. On the other hand, Earth's history represents only a single datum among the possible histories of the $\sim 10^{10}$ Earth-like planets orbiting the habitable

**Corresponding author*: amedeo.balbi@roma2.infn.it

†*Corresponding author*: claudio.grimaldi@epfl.ch

Anamaria Berea (ed.) Technosignatures for Detecting Intelligent Life in Our Universe: A Research Companion, (127–148) © 2022 Scrivener Publishing LLC

zones of sun-like stars in the Milky Way galaxy [8.14][8.34][8.47]. As a matter of fact, the rate of abiogenesis remains utterly unknown even for the simplest forms of life, let alone the technogenesis rate of hypothetical exo-civilizations, at least as long as a second example of life (either intelligent or not) is found [8.52][8.5].

This state of ignorance has not hampered the proliferation of a vast literature discussing the probability of technological life emerging beyond Earth, referring in particular to the so-called Fermi-Hart paradox [8.29][8.4][8.56][8.7][8.18] and the closely related Drake equation [8.12][8.13]. The Fermi-Hart paradox is an argument contrasting the lack of any evidence that the Earth has ever been visited by space-faring civilizations with the relatively short timescales needed to colonize the entire galaxy. Several solutions to this paradox have been proposed along the years, ranging from astrophysical adverse events (like, e.g., supernovae and gamma-ray bursts) preventing the development of advanced space-faring civilizations [8.9][8.30][8.1] to the extreme rarity of complex life in the universe [8.29][8.55].

A broader formulation of the Fermi-Hart argument refers to the lack of any sign of intelligence beyond Earth that could be remotely detected. While traditionally the search for technological life elsewhere in the universe was mostly restricted to artificial electromagnetic emissions (especially radio) from other planetary systems, the scope of the investigation is in principle much vaster, as it can include any technosignature, i.e., any detectable sign that can be produced by technological activity—such as, for example, large-scale engineering [8.15][8.57], pollution, alteration in planetary atmosphere and climate, artificial spectral signatures [8.40][8.43][8.49][8.38][8.36][8.3], laser bursts from lightsail propulsion [8.45][8.40], satellite belts [8.51], and so on. This only deepens the need to address the Fermi-Hart argument, but it also complicates the task of producing theoretical prediction of the type and frequency of possible technosignatures. Although each of these possible signals has different causes and comes with its own specific features, the only way we can gain knowledge of them is by remote investigation through electromagnetic radiation. This is why, in this contribution, we will often refer, generically, to "emissions", without getting into the details of the specific technosignature that caused them.

In parallel to theoretical approaches, active searches for extraterrestrial technosignatures have been pursued during the last sixty years, starting with the *Nature* paper of [8.10], followed by the project Ozma of Frank Drake [8.12] and his eponymous equation. Since then, a number of privately and publicly funded search projects have checkered the history of what is commonly known under the acronym SETI: the search for extraterrestrial intelligence. SETI has been pursued primarily in the United States, but also in the former USSR and post-Soviet Russia [8.23]. These initiatives have found no evidences of technologically advanced civilizations emitting, either intentionally or not, in the radio [8.53][8.16] and optical domains [8.54]. Equally fruitless have been the searches for infrared emissions from the waste heat of putative megastructures [8.25], such as Dyson spheres [8.15] or other artifacts harvesting energy on a planetary, stellar, or galactic scale [8.33]. When coupled with the sheer immensity of the parameter search space, however, the absence of detection does not warrant any firm conclusion about the existence of potentially detectable exo-civilizations [8.53]. This point has been recently emphasized by [8.58] who compare the tiny fraction of the parameter space explored so far to the ratio of the volume of a small swimming pool to that of the Earth's oceans.

Despite this frustrating state of ignorance, the discovery that Earth-like exoplanets exist in abundance throughout the galaxy and the exponential advances in detector technologies and computing power have recently boosted dramatically the search for technosignatures. The privately funded "Breakthrough Listen" program started in 2016 is the largest and most comprehensive search ever [8.32][8.16]. It aims to probe a million nearby stars, the plane of the Milky Way, and 100 nearby galaxies over a period of ten years. In June 2019, "Breakthrough Listen" announced the largest-ever release of SETI data, comprising radio and optical spectra of over 1,700 nearby stars [8.48]. Furthermore, in April 2018 the US Congress mandated that NASA begin supporting the scientific search for technosignatures as part of the agency's search for life [8.46], while in March 2019 eleven white papers on SETI were proposed for the Astronomy and Astrophysics Decadal Survey 2020-2030 of the US National Academy of Sciences. The search for radio technosignatures is also planned to be part of the "Cradle of Life" program of the future Square Kilometre Array telescope [8.37] [8.50].

Surprisingly, despite this recent burst of initiatives and the decades-old history of SETI, there have been only limited efforts in developing full physical-statistical descriptions of the electromagnetic technosignatures that could possibly populate the galaxy. It is significant that, even if ongoing and future SETI efforts will search the galaxy and the local universe with ever-increasing bandwidth, sky coverage, and sensitivity, no statistical tools have yet been fully deployed to analyze systematically the prospects of either the detection or the non-detection of electromagnetic technosignatures.

8.2 General Issues

The search for technosignatures presents several conceptual and methodological problems [8.20]. The most outstanding are 1) the lack of a definite and accepted predictive framework to plan and interpret observations, and 2) the possibility of statistical biases that can affect the inference from existing data. The two problems are related, since most predictive models produced so far are probabilistic in nature, relying on some hypothesis on the occurrence of the various astrophysical and biological processes required for the emergence of technological exo-civilizations. However, since no prior knowledge is available on the frequency of such occurrences, the standard strategy was the adoption of non-informative distributions, or the assumption of stationarity of the underlying statistical processes, or the reliance on the "typicality" or "mediocrity" of what happened on Earth. All of these assumptions, which are hard to justify empirically, are often uncritically adopted and left implicit in the discussion of results.

A further complication is that, when dealing with signals produced by technological activity, deterministic laws may be an inadequate predictive tool, since agency and purpose plays a crucial role: think, for example, of the huge difference between biological and cultural evolution, and of the difficulties involved in forecasting the trajectory of the latter with respect to the former. This implies, among other things, that it may be impossible to decide what can be commonly expected. In fact, it may be in the very nature of technosignatures searches to focus on anomalies and outliers, rather than on "average" or "normal". This is why a special care has to be taken in making explicit the statistical assumptions underlying any analysis of technosignatures.

The theoretical tool traditionally used to frame the discussion around technosignatures is the Drake equation, which in its most basic form can be written as:

$$N_D = \Gamma \langle L \rangle,$$

The equation estimates the mean number of *actively* emitting civilizations, N_D, from the product between the birthrate Γ and the average longevity $\langle L \rangle$ of the emission processes [8.12][8.13]. In Drake's original formulation, Γ was expressed as a product of probabilities of events thought to be conducive to the emergence of life and to the development of technological exo-civilizations:

$$\Gamma = R_{\star} f_p n_e f_l f_i f_c$$

where $R_{\star}$ is the star formation rate, f_p the fraction of star hosting planetary systems, n_e the number of habitable planets per planetary system, f_l the fraction of habitable planets developing life, f_i the fraction of inhabited planets where intelligent species appear, f_c the fraction of planets with intelligent species making the transition to technology that enables interstellar communication of some sort. This establishes therefore a direct link to the arguments pertaining to the Fermi-Hart paradox. Not surprisingly, lacking any hint about the actual value of Γ and $\bar{L}$, estimates of N_D span several orders of magnitude, even for the Milky Way galaxy alone.

While much of the existing scientific literature on the Drake equation has focused on guessing the numerical value of its various factors, very little work has been devoted to its statistical structure and on the underlying assumptions. An especially troublesome aspect, which has received little attention, is the lack of an explicit temporal dependence in the equation, although clearly there is no guarantee that its factors remained constant in time [8.6]. In particular, the average appearance rate in the Drake equation, Γ is usually estimated by guessing its present value: however, not only the rate of star formation $R_{\star}$ certainly changed over cosmic history, but also the other probabilistic factors involved might have had different values at different epochs. Also, the distribution of exo-civilizations might not be spatially uniform in the volume under consideration. Many authors have discussed how the probability of the appearance of life may vary over the course of cosmic history [8.44][8.11] and how life might not be spread uniformly in the galaxy [8.24][8.39]. However, such spatiotemporal dependencies are not usually incorporated in the Drake equation in its standard form. Indeed, the Drake number N_D is equally quoted in the literature as the time-averaged number of communicating civilizations existing in the galaxy, or their present-day number, implicitly assuming that the number has remained stationary over the history of the universe.

In fact, though not often stated, the Drake equation deals with average quantities. This is apparent when one notices that the equation is just a specific example of a remarkable result of queueing theory, the so-called Little's law [8.41], that relates the long-term average number of items in a stationary system (for example customers waiting to be served) with the long-term average effective arrival rate of the items multiplied by the average time that an item spends in the system. In our case, the system is the Earth's location (or more generally a volume of space around it) and the average arrival rate is simply the appearance rate of emission processes released by technological civilizations. Note that in interpreting the Drake equation in terms of Little's law, we are supposing that the emissions released from any location in the Milky Way inevitably intersects the Earth during their travel through the galaxy. This is possible either for isotropic emissions or for anisotropic signals (like, e.g., beamed emissions) directed towards Earth. In the following we shall see that, indeed, Drake's number N_D does not necessarily coincide with the average number of emission processes intersecting Earth. Another important point is that Little's result does not depend on the statistical distribution of the variables involved, but does assume that the underlying random process is stationary [8.42]. Thus, it cannot treat processes whose arrival rate is expected to vary in time, although there have been attempts to derive time-varying versions of the law [8.35].

It may seem that the assumption of stationarity may be made reasonable by appealing to some "principle of mediocrity", i.e., the presumption that our location in time and space is typical. However, at least with respect to time, this is almost certainly not true. In fact, in an evolving universe, typicality in time cannot be a well-defined assumption [8.8]. More generally, the uncritical application of a principle of mediocrity is precisely one of the potential sources of problems in any attempt to assess the frequency of intelligent life in the universe. There is no way of telling whether the only available datum (our own existence) reflects the commonality of intelligence in the universe or, instead, is a manifestation of survivor bias. At the very least, we should regard with extreme caution any attempt of inferring general statistical conclusions from a single observation.

Another important point to keep in mind is that, in general, we are not simply interested in the estimate of the number of technological civilizations that *ever* existed in the universe, although this has been done in some cases [8.21]. Rather, we look for empirically detectable technosignatures. This means that only a fraction of all the exo-civilizations that (possibly) existed can actually be observed by us, today. This is directly related to the spatiotemporal distribution of the emitters, since they have to lie in our past light-cone. We will explore the statistical consequences of this fact in more detail later, but for now let us consider a single exo-civilization located at distance d from Earth that started being visible at time t in the past, and remained visible for a time L. As shown in Figure 8.1, we will be able to observe this technosignature if and only if

$$t - L \leq d/c \leq t, \tag{8.1}$$

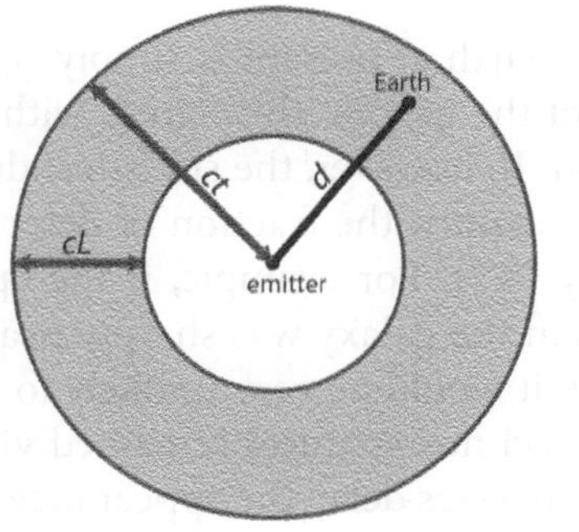

Figure 8.1 Schematic illustration of the causal constraint of Eq. (8.1). The annular region represents the volume filled by the electromagnetic radiations released by an emitter. The outer radius of the annulus is ct, where t is the time from present at which the emitter started transmitting, and L is the time duration of the emission. The annular region will intersect the Earth if the distance d from the emitter satisfies Eq. (8.1).

where c is the speed of light. This apparently obvious causal constraint has interesting implications, because it introduces a correlation between two otherwise independent timescales, i.e., the time of appearance of the technosignature, t, and its duration L. Note that t, in principle, can assume any value between the present ($t = 0$ years) and the age of the most ancient stars in our galaxy ($t \sim 10^{10}$ years), while there is no estimate of a plausible range for L(although it certainly cannot exceed the age of the galaxy). However, because the quantity d/c can be at most $\sim 10^5$ years for stars within our galaxy, equation (8.1) dictates that only technosignatures having $L \sim t$ can actually be observed. The required match between t and L gets even more precise for locations closer to Earth. In practice, this means that detectable technosignatures must either be very long-lived (if their appearance t was far back in the past) or almost exactly coeval to us.

It must be noted that equation (8.1) must hold for any single exo-civilizations that we can detect. So it does not rely, *per se*, on statistical models of the distribution of technosignatures. When we extend the causal constraint to an ensemble of possible exo-civilizations at various distances, however, the causal constraint acts as a filter, and the resulting number of detections depends on the temporal and spatial distribution of sources. If N_{tot} is the number of technosignatures that ever existed in the galaxy, the fraction that is in causal contact with us can be defined as $f_{det} = N_{det}/N_{tot}$ where the detectable number of technosignatures N_{det} are those satisfying the causal constraint (8.1). If t is uniformly distributed over the history of the galaxy (as usually assumed in the standard estimates based on the Drake Equation), then there is no preferred value for the epoch of appearance of intelligent life. In this case, the fraction of technosignatures that can be detected from Earth at present is simply $f_{det} = \overline{L} / T_G$, where $T_G \sim 10^{10}$ years is the age of the galaxy. This agrees with the standard estimate of the Drake Equation. If, however, the statistical distribution of t departs significantly from uniformity, the fraction of detectable technosignatures will vary significantly [8.2]. For example, if the epoch of appearance of technological species in the galaxy was sharply peaked around, say, a few billion years ago, then it would be very unlikely to catch signals from any of them, unless their technosignatures remained visible for $L \sim 10^9$ years. Conversely, if some processes delay the appearance of technological civilizations at later stages in the evolution of the Milky Way, a larger fraction of short-lived technosignatures could be in causal contact with us at present.

For the same reason, the assumed longevity of technosignatures plays a crucial role in assessing the probability of their detection. The average value of L may be inadequate to fully capture the complexity of real-life scenarios, especially if the distribution of L is fat-tailed or heavily skewed.

These considerations can be made more quantitative, and should be accounted for in the analysis of existing and future observations, especially to interpret negative results. This is the argument of the following sections.

8.3 Emission Processes

In defining our statistical model, we assume that hypothetical artificial emitters are located at random site $\mathbf{r}$ relative to the galactic center and that they are statistically independent of each other. In the following, we employ the generic term "emission process" to indicate an artificial EM radiation of any wavelength and power spectrum that is emitted either continuously or not during a time duration L. We shall, however, make a distinction between isotropic processes, that is, radiating in all directions, and those that are anisotropic, such as randomly directed beam-like signals or beacons sweeping across space, by labelling each type of emission by the index i. Furthermore, we neglect scattering and absorption processes by the interstellar medium and assume that the radiations propagate unperturbed throughout the galaxy at the speed of light c.

We introduce the rate of appearance per unit volume for emitters of type i, $\gamma_i(\mathbf{r}, t)$, defined so that $\gamma_i(\mathbf{r}, t)\, d\mathbf{r}dt$ gives the expected number of i-emitters within a volume element $d\mathbf{r}$ about $\mathbf{r}$ that started emitting within a time interval dt centered at a time t before present. Since we are interested in emitters located in the Milky Way, we take the emitter rate of emergence to vanish outside the Galaxy, so that integrating γ_i ($\mathbf{r}$, t) over $\mathbf{r}$ gives the birthrate of the emission processes of type i in the entire Milky Way:

$$\Gamma_i(t) = \int d\mathbf{r}\, \gamma_i(\mathbf{r}, t),$$

while

$$\Gamma(t) = \sum_i \Gamma_i(t)$$

is the total emission rate. Finally, we associate to all processes of type i a common probability distribution function (PDF) of the longevity, denoted ρ_i (L), such that

$$\int_0^\infty dL\, \rho_i(L) L^n = \langle L^n \rangle_i$$

is the n-moment of the longevity of the processes of type i.

8.3.1 Drake's N_D

As we mentioned in Sec. 8.2, the scientific literature on the Drake equation equally interprets N_D as the time-averaged number of exo-civilizations existing in the Milky Way, or their mean number at the present time. Here, we show that a rigorous and unambiguous definition of N_D refers to the average number of emitters that are currently transmitting, regardless of whether or not their emission processes intersect the Earth. In this way, we explicitly distinguish N_D from the number of emissions that could be possibly detected on Earth, and that are therefore necessarily subject to the causal constraint given in Eq. (8.1).

We start by noticing that for any galactic emitter that started an emission process at a time t in the past, the necessary condition that at present time $t = 0$ the emitter is still active is that the time elapsed since its birth is shorter than the emission longevity, that is $t \leq L$. The time integral between $t = 0$ and $t = L$ of γ_i ($\mathbf{r}$, t) gives therefore the number of active i-emitters within $d\mathbf{r}$ having longevity L. Extending to the entire Galaxy and over all allowed longevities yields the total number of active emitters (Drake's number):

$$N_D^i = \int_0^\infty dL\, \rho_i(L) \int_0^L dt \int d\mathbf{r}\, \gamma_i(\mathbf{r},t) = \int_0^\infty dL\, \rho_i(L) \int_0^L dt\, \Gamma_i(t) \tag{8.2}$$

Under the assumption that the time dependence of the galactic birthrate can be neglected, $\Gamma_i(t) \approx \Gamma_i$, the time integration in Eq. (8.2) yields the compact form of the Drake equation for active i-emitters:

$$N_D^i = \Gamma_i \langle L \rangle_i . \tag{8.3}$$

Introducing $x_i = \Gamma_i/\Gamma$, with $\Gamma = \sum_i \Gamma_i$, and the total average longevity $\langle L \rangle = \sum_i x_i \langle L \rangle_i$, the sum over i of (3) reduces to the Drake equation $N_D = \Gamma\langle L \rangle$.

Let us now analyze the conditions of validity of the Drake equation if the hypothesis of a time independent rate of emission is relaxed. For the sake of simplicity, we adopt for the rate density the separable ansatz $\gamma_i(\mathbf{r}, t) = \gamma_i(\mathbf{r}) f(t)$, which treats the time evolution as being independent of the emitter position. Instead of specifying a model for the full functional dependence of $f(t)$, here we consider the Taylor expansion of $f(t)$ and restrict the analysis to the first order in t: $f(t) = 1 \pm t/\tau$, where τ is a characteristic timescale and the + sign (−sign) corresponds to a higher (lower) birthrate in the past compared to the present value. After performing the integrals in Eq. (8.2) we find:

$$N_D^i = \Gamma_i \langle L \rangle_i \pm \frac{\Gamma_i}{2\tau} \langle L^2 \rangle_i , \tag{8.4}$$

from which we see that the validity of Eq. (8.3) breaks down when $\tau^{-1} \langle L^2 \rangle / \langle L \rangle_i$ is not a small quantity compared to the unity. For longevity distributions with small variance, the validity of the Drake equation (8.3) therefore holds true as long as $\langle L \rangle << \tau$. For example, under the assumption that the time scale over which $\Gamma_i(t)$ varies most is comparable to that of the formation rate of terrestrial planets in the Milky Way, i.e., $\tau \approx 1$ Gyr, an upper bound on $\langle L \rangle_i$ of a few tens of Myr is largely sufficient to make the last term in Eq. (8.4) negligible. However, if instead we make the hypothesis that $\tau \approx \langle L \rangle_i$, or that $\langle L^2 \rangle_i / \langle L \rangle_i >> \langle L \rangle_i$, as it would be the case for a weakly decaying (i.e., power law) tail of $\rho_L(L)$ at large L, the temporal correction to the usual Drake equation should be definitely taken into account.

8.3.2 Average Number of Emission Processes at Earth

Having shown that the Drake's number N_D gives the average number of currently active emitters, and that the Drake equation (8.2) applies only for a stationary birthrate of emitters, we now move to analyze another important statistical indicator: the average number $\bar{k}$ of emission processes crossing Earth, which is the quantity of direct interest in the search for technosignatures.

Let us start by showing that $\bar{k}$, even if it is sometimes confused with the Drake number, it actually coincides with N_D only by assuming a constant birthrate of isotropic emissions. To see this, we take an emitter located at $\mathbf{r}$ that started emitting at a time t in the past an isotropic signal for a duration L. If the process is continuous, at the present time the region of space filled by the EM radiations is a spherical concentric shell of outer radius ct and thickness cL, centered on $\mathbf{r}$. The condition that at $t = 0$ the shell intersects the Earth is fulfilled by the requirement (1), where $d = |\mathbf{r} - \mathbf{r}_E|$ and $\mathbf{r}_E$ is the vector position of the Earth. If $\gamma_{iso}(\mathbf{r}, t)$ (where the subscript "iso" stands for isotropic) is the process birthrate per unit volume and $\rho_{iso}(L)$ is the PDF of the longevity, the average number of spherical shells at Earth is obtained by marginalizing the condition (8.1) over $\mathbf{r}$, L, and t:

$$\bar{k}_{iso} = \int dL\, \rho_{iso}(L) \int d\boldsymbol{r} \int_{|\boldsymbol{r}-\boldsymbol{r}_E|/c}^{|\boldsymbol{r}-\boldsymbol{r}_E|/c+L} dt\, \gamma_{iso}(\boldsymbol{r},t). \tag{8.5}$$

In the case in which the temporal dependence of the birthrate can be neglected, $\gamma_{iso}(\mathbf{r}, t) \simeq \gamma_{iso}(\mathbf{r})$, Eq. (8.5) reduces to:

$$\bar{k}_{iso} = \int dL\, \rho_{iso}(L) L \int d\boldsymbol{r}\, \gamma_{iso}(\boldsymbol{r}) = \Gamma_{iso} \langle L \rangle_{iso}, \tag{8.6}$$

which coincides with the Drake number N_D^{iso} relative to isotropic emission processes [8.27]. Note that since we have taken a time independent birthrate, $\mathbf{r}_E$ has dropped off Eq. (8.6), meaning that in the steady state $\bar{k}_{iso}$ actually gives the mean number of emissions crossing any given point in the galaxy.

Let us see now how $\bar{k}_{iso}$ is affected by a weak temporal dependence of $\gamma_{iso}(\mathbf{r}, t)$. We take the ansatz $\gamma_{iso}(\mathbf{r}, t) = \gamma_{iso}(\mathbf{r})(1 \pm t/\tau)$ and perform the integration over t in Eq. (8.5):

$$\begin{aligned}\bar{k}_{iso} &= \int dL\, \rho_{iso}(L) \int d\boldsymbol{r}\, \gamma_{iso}(\boldsymbol{r}) \left(L \pm \frac{L^2}{2\tau} \pm \frac{|\boldsymbol{r}-\boldsymbol{r}_E|}{c\tau} \right) \\ &= \Gamma_{iso} \langle L \rangle_{iso} \pm \frac{\langle L^2 \rangle_{iso}}{2\tau} \pm \frac{\langle L \rangle_{iso}}{c\tau} \int d\boldsymbol{r}\, \gamma_{iso}(\boldsymbol{r}) |\boldsymbol{r}-\boldsymbol{r}_E|. \end{aligned} \tag{8.7}$$

The first two terms in the second equation corresponds to the modified Drake's number (8.4), so that to the first order in $\tau^{-1}\langle L^2\rangle_{iso}/\langle L\rangle_{iso}$ Eq. (8.7) can be rewritten as

$$\bar{k}_{iso} \simeq N_D^{iso}\left[1 \pm \frac{1}{c\tau\Gamma_{iso}}\int d\boldsymbol{r}\,\gamma_{iso}(\boldsymbol{r})|\boldsymbol{r}-\boldsymbol{r}_E|\right], \tag{8.8}$$

showing that $\bar{k}_{iso}$ no longer formally coincides with Drake's number N_D^{iso}, and that now $\bar{k}_{iso}$ depends explicitly on the vector position of the Earth. To evaluate explicitly Eq. (8.8), we take an axisymmetric approximation for $\gamma_{iso}(\mathbf{r})$ of the form:

$$\gamma_{iso}(\boldsymbol{r}) = \Gamma_{iso}\frac{e^{-\frac{r}{r_s}}e^{-\frac{|z|}{z_s}}}{4\pi r_s^2 z_s}, \tag{8.9}$$

which using $r_s = 8.15$ kly and $z_s = 0.52$ kly basically reproduces the star distribution in the thin disk of the Milky Way. After numerical integration of Eq. (8.8) we find:

$$\bar{k}_{iso} \simeq N_D^{iso}\left(1 \pm 1.15\frac{r_E}{c\tau}\right)$$

In the case of isotropic emissions, therefore, $\bar{k}_{iso}$ deviates from the corresponding Drake's number N_D^{iso} as soon as the time-scale τ is comparable to $r_E/c \simeq 30'000$ yr, which is the time it takes to a photon to travel from the center of the galaxy to the Earth. Unless the rate of emission did not experience substantial changes in the last 30'000 years, we can thus safely assume an effective equivalence between $\bar{k}_{iso}$ and N_D^{iso}.

Even for a formally time-independent birthrate, this equivalence is however limited to the case of isotropic emissions, and it inevitably breaks down when we consider anisotropic processes. This is simply due to the fact that the Drake's number enumerates the active emission processes regardless of their directionality, while only the fraction of anisotropic emissions that are directed towards Earth can contribute to $\bar{k}$.

This can be seen more quantitatively by considering the simple case of an emitter located at **r** that transmits since a time t before present a conical beam of aperture α_0 [8.19][8.26]. During the entire lifetime L of the emission, the beam axis is held oriented along the direction of a unit vector $\hat{n}$. The region of space filled by the radiation is the intersection between a cone of apex at **r** and a spherical shell centred on the cone apex with outer radius ct and thickness cL. If we take a constant birthrate of beams having angular aperture α and beam axis orientations distributed uniformly over three dimensions (3D), the resulting mean number of beams crossing Earth, $\bar{k}_{\text{ani}}$, is weighted by the fractional solid angle subtended by the beam, that is, $\frac{1}{2}[1-\cos(\alpha/2)]$. For beam directions distributed uniformly over the two-dimensional (2D) galactic plane the fractional solid angle becomes $\alpha/2\pi$. In the limit of narrow beams therefore

$$\bar{k}_{ani} \simeq \begin{cases} \dfrac{\alpha^2}{16}\Gamma_{ani}\langle L\rangle_{ani}, & \text{3D beams,} \\[2ex] \dfrac{\alpha}{2\pi}\Gamma_{ani}\langle L\rangle_{ani}, & \text{2D beams,} \end{cases} \tag{8.10}$$

which shows that $\bar{k}_{\text{ani}}$ can be orders of magnitude smaller than the corresponding Drake's number for beam-like signals $N_D^{\text{ani}} = \Gamma_{ani}\langle L\rangle$. Taking for example beam apertures comparable to that of the Arecibo radar ($\sim 2' \simeq 6\times10^{-4}$ rad) Eq. (8.10) yields $\bar{k}_{\text{ani}}/N_D^{\text{ani}} \sim 2\times10^{-8}$ and $\sim 10^{-4}$ for 3D and 2D random beams, respectively. Even smaller values of $\bar{k}_{\text{ani}}/N_D^{\text{ani}}$ are obtained by assuming optical or infrared laser emission of apertures under an arcsecond [8.31][8.54].

8.4 Bayesian Inference from Non-Detection and Detection

The statistical analysis of the previous sections has clarified the importance of distinguishing between observable quantities, such as the number of emissions crossing Earth, and non-observable ones, as the number of active emitters N_D. We have shown that although being governed by the signal birthrate and longevity, these two statistical indicators generally differ from each other when the assumptions of stationary birthrates and signal isotropy are relaxed. We now move to briefly discuss what could be empirically learned from observations and, in particular, what are the consequences of either detecting or not detecting EM technosignatures in current and future searches.

We presently lack either philosophical or empirically based arguments to estimating, even at the order of magnitude level, the galactic population of signals impinging upon Earth. However, ongoing and planned large-scale search initiatives give hope for the future of SETI, and the following years and decades will likely see the search for technosignatures evolving to unprecedented levels. In particular, the advent in the last decades of aperture arrays instruments and the development of large-scale telescope arrays with large field-of-view offer significant advantages compared to single-dish radio telescopes, as larger portions of the sky can be searched in less time and with competitive sensitivities [8.50][8.22].

The data gathered by future wide field-of-view SETI searches or, ideally, by all-sky SETI surveys can enable probabilistic inferences about the possible population of extraterrestrial signals in the Milky Way. In particular, $\bar{k}$ is the quantity that can be in principle inferred by observations and that combines other variables that cannot be directly measured, such as the temporal longevity of the signal process, the rate of emissions, the signal directionality, etc. In statistical terms, $\bar{k}$ is an average over the possible configurations (realizations) of the signals crossing our planet and, owing to the statistical independence of the emitters, it can be expressed as a linear combination of the different types of emission processes: $\bar{k} = \sum_i \bar{k}_i$. For example, assuming that the galaxy is populated only by isotropic and beamed-like signals directed uniformly over the 3D space, the mean number of them that are crossing Earth becomes

$$\bar{k} = \bar{k}_{\text{iso}} + \bar{k}_{\text{ani}} = \Gamma_{\text{iso}} \langle L \rangle_{\text{iso}} + \frac{\alpha^2}{16} \Gamma_{\text{ani}} \langle L \rangle_{\text{ani}} .$$

In principle, the region of space searched by an all-sky SETI survey is a sphere centered at Earth and of radius $R = \sqrt{L_E / 4\pi S_{\text{min}}}$, where L_E is the equivalent isotropic radiated power (EIRP), or effective luminosity, of an emitter, and S_{min} is the detection threshold defined as the minimum EM flux that can be detected by the survey. Either non-detection or the detection of a signal within R represents a datum, D, that can be used to infer $\bar{k}$ by means of Bayes' theorem:

$$p(\bar{k} \mid D) = \frac{P(D \mid \bar{k}) p(\bar{k})}{P(D)}, \tag{8.11}$$

where $p(\bar{k})$ is the prior probability distribution function (PDF) we ascribe to $\bar{k}$ before having gathered the datum $D, P(D\,|\,\bar{k})$ is the likelihood function, defined as the conditional probability of observing D given $\bar{k}$, and $P(D)=\int d\bar{k}p(d\,|\,\bar{k})p(\bar{k})$. Assuming suitable initial hypotheses about $p(\bar{k})$, Eq. (8.11) yields the posterior PDF of $\bar{k}, p(\bar{k}\,|\,D)$ resulting from the data gathered by the survey.

There are three general types of data that can be gathered by observations within a distance R from Earth: the non-detection (D_0), the detection of exactly one signal (D_1), and the detection of at least one signal ($D_{\geq 1}$) (this latter case corresponds to the event of detection within a fraction of the search volume). Under the assumption of a Poisson distribution of signals, the corresponding likelihood functions are given by [8.28][8.17]:

$$P(\bar{k}\,|\,D_0)=e^{-\pi(R)\bar{k}}, \quad P(\bar{k}\,|\,D_1)=\pi(R)\bar{k}e^{-\pi(R)\bar{k}}, \quad P(\bar{k}\,|\,D_{\geq 1})=1-e^{-\pi(R)\bar{k}},$$

where $\pi\,(R)$ is the probability of an emitter being within a distance R from the Earth. We adopt the hypothesis that the spatial distribution of the emitters does not depend on the specific type of the emissions. We further assume that it follows the functional form of Eq. (8.9). Hence:

$$\pi(R)=\int d\mathbf{r}\frac{e^{-r/r_s}e^{-|z|/z_s}}{4\pi r_s^2 z_s}\theta(R-|\mathbf{r}-\mathbf{r}_E\,|),$$

where $\theta\,(x)$ is the Heaviside step function.

Since, as discussed above, we ignore even the order of magnitude of the signals crossing Earth, a suitable choice for the initial PDF of $\bar{k}$ (denoted also the prior PDF) is given by a distribution that is uniform in $\log(\bar{k})$, which gives equal weight to all orders of magnitude [8.52]. The present lack of detection can be taken into account by introducing an upper cut-off in $\bar{k}$, which can be determined by considering a prior detection threshold, $S_{\rm min}^{\rm prior}$, representative of previous all-sky searches within a given frequency range.

Figure 8.2 shows the posterior probability that $\bar{k}\geq 1$ (top panel) and $\bar{k}\geq 100$ (bottom panel) resulting from non-detection or the detection of a signal within a distance R from Earth, assuming that the emitters have EIRP values equivalent to that of the Arecibo radar ($L_{\rm Arecibo}$ = 2 × 10^{13} W) and that they transmit within a range comprised between 1 GHz and 2 GHz. The prior probability (dashed lines) have been obtained by assuming $S_{\rm min}^{\rm prior}=10^{-23}$ Wm^{-2} for this frequency range [8.28]. The probability of

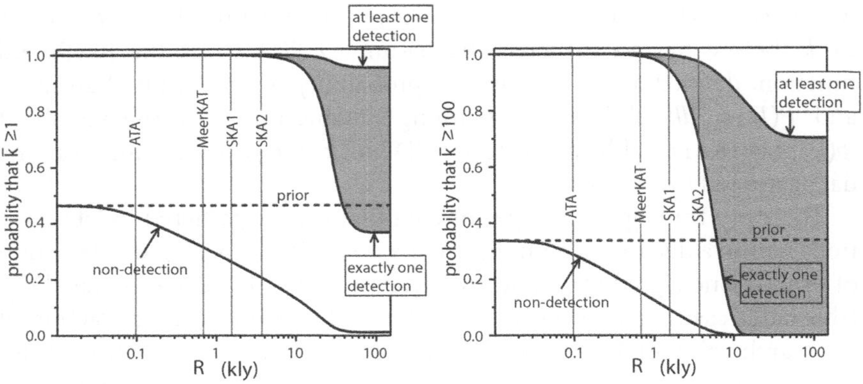

Figure 8.2 Posterior probability that $\bar{k} \geq 1$ (left) and $\bar{k} \geq 100$ (right) resulting from non-detection, at least one detection, and exactly one detection of a signal within a distance R from Earth. The vertical lines are the R values corresponding to the distances accessible to the ATA, MeerKAT, SKA1, and SKA2 telescopes assuming that the emitters have EIRP values equal to that of the Arecibo radar transmitter.

detection (shaded area) encompasses the values that the posterior probability can take if at least one or exactly one signal is detected within a sphere of radius R. The vertical lines represent the values of R that are accessible, for $L_E = L_{\text{Arecibo}}$, by the ATA, MeerKAT, SKA1, and SKA2 telescopes, assuming that their SEFD values are respectively 664 Jy, 8.6 Jy, 1.7 Jy, and 0.3 Jy, with a signal-to-noise ratio of 15, an integration time of 10 min, and a receiver channel bandwidth of 0.5 Hz.

The results shown in Figure 8.2 highlight the informative value that a Bayesian analysis of all-sky SETI data potentially gathered by current and future telescope arrays can bring. Non-detecting signals by the MeerKAT, SKA1, and SKA2 telescopes essentially halves the chances that typically more than one signal, emitted with an EIRP equivalent to that of the Arecibo radar, crosses the Earth (left panel). The probability that $\bar{k} \geq 100$ drops to only about 3% if the SKA2 telescope detects no signals (right panel).

Conversely, the detection of even a single signal would imply a dramatic revision of our prior hypothesis. One would be led to conclude with almost certainty that other signals than the one already detected are crossing our planet ($\bar{k} > 1$, left panel), while $\bar{k} \geq 100$ is expected with a probability above 70% (right panel). Interestingly, from the analysis of the previous section it turns out that it is unlikely that above 100 signals are all narrow beams, as this implies for example that the corresponding Drake's number, $N_D^{\text{ani}} = \Gamma_{\text{ani}} \langle L \rangle$ for Arecibo-like beams ($\alpha \simeq 2'$) would exceed $16\bar{k}_{\text{ani}} / \alpha^2 \sim 5 \times 10^9$, a number comparable to the population of habitable

Earth-like planets in the Milky Way. This teaches us that Bayesian inference analysis can in principle tell us more than just the possible population of detectable technosignatures, and that a statistical methodology like the one here outlined should be further developed in order to fully understand the implications of non-detection or of a possible, future technosignature detection.

References

[8.1] Annis, J. (1999) An astrophysical explanation for the "great silence". *Journal of the British Interplanetary Society* 52, 19-22.

[8.2] Balbi, A. (2018) The Impact of the Temporal Distribution of Communicating Civilizations on Their Detectability. *Astrobiology* 18, 54-58.

[8.3] Berdyugina, S. V., and Kuhn, J. R. (2019) Surface Imaging of Proxima b and Other Exoplanets: Albedo Maps, Biosignatures, and Technosignatures. *Journal of Astronomy* 158, #246.

[8.4] Brin, D. (1983) The Great Silence - the Controversy Concerning Extraterrestrial Intelligent Life. *Quarterly Journal of the Royal Astronomical Society* 24, 283-309.

[8.5] Chen, J. and Kipping, D. (2018) On the rate of abiogenesis from a Bayesian informatics perspective. *Astrobiology* 18, 1574-1584.

[8.6] Ćirković, M. M. (2004) The Temporal Aspect of the Drake Equation and SETI. *Astrobiology* 4, 225-231.

[8.7] Ćirković, M. M. (2018) *The Great Silence: The Science and Philosophy of Fermi's Paradox*. Oxford University Press, Oxford, UK.

[8.8] Ćirković, M. M., and Balbi, A. (2020) Copernicanism and the typicality in time. International Journal of Astrobiology, 19, 101-109.

[8.9] Clarke, J. N. (1981) Extraterrestrial intelligence and galactic nuclear activity. *Icarus* 46, 94-96.

[8.10] Cocconi, G. and Morrison, P. (1959) Searching for Interstellar Communications. *Nature* 184, 844-846.

[8.11] Dayal, P., Ward, M., and Cockell, C. (2016) The habitability of the Universe through 13 billion years of cosmic time. *eprint arXiv:1606.09224.*

[8.12] Drake, F. D. (1961) Project Ozma. *Physics Today* 14, 40.

[8.13] Drake, F. D., 1965, The radio search for intelligent extraterrestrial life. In *Current Aspects of Exobiology*. Mamikunian, G. & Briggs, M. H. (eds.) Pergamon Press, New York, USA. pp. 323-345.

[8.14] Dressing, C. D., and Charbonneau, D. (2013) The occurrence rate of small planets around small stars. *The Astrophysical Journal* 767, #95.

[8.15] Dyson, F. J. (1960) Search for Artificial Stellar Sources of Infrared Radiation. *Science* 131, 1667-1668.

[8.16] Enriquez, J. E., Siemion, A., Foster, G., Gajjar, V., Hellbourg, G., Hickish, J., Isaacson, H., Price, D. C., Croft, S., DeBoer, D., Lebofsky, M., MacMahon, D. and Werthimer, D. (2017) The Breakthrough Listen Search for Intelligent Life: 1.1-1.9 GHz Observations of 692 Nearby Stars. *The Astrophysical Journal* 849, #104.

[8.17] Flodin, M., (2019) *Determining upper limits on galactic ETI civilizations transmitting continuous beacon signals in the radio spectrum* (Dissertation). Retrieved from http://urn.kb.se/resolve?urn=urn:nbn:se:kth:diva-266824.

[8.18] Forgan, D. H. (2019) *Solving Fermi's Paradox*. Cambridge University Press, Cambridge, UK.

[8.19] Forgan, D. (2014) Can Collimated Extraterrestrial Signals be Intercepted? *Journal of the British Interplanetary Society* 67, 232-236.

[8.20] Forgan, D., Wright, J., Tarter, J., Korpela, E., Siemion, A., Almár, I. and Piotelat, E. (2019) Rio 2.0: revising the Rio scale for SETI detections. *International Journal of Astrobiology* 18, 336-344.

[8.21] Frank, A. and Sullivan, W. T., III (2016) A New Empirical Constraint on the Prevalence of Technological Species in the Universe. *Astrobiology* 16, 359-362.

[8.22] Garrett, M. A., Siemion, A., and Cappellen, W. A. (2017) All-sky radio SETI. MeerKAT Science: On the Pathway to the SKA (MeerKAT2016), eds Taylor P, Camilo F, Leeuw L, Moodley K, International School for Advanced Studies, Trieste, Italy, 277, #020.

[8.23] Gindilis, L. M. and Gurvits, L. I. (2019) SETI in Russia, USSR and the post-Soviet space: a century of research. *Acta Astronautica* 162, 1-13.

[8.24] Gonzalez, G. (2005) Habitable Zones in the Universe. *Origins of life and evolution of biospheres* 35, 555-606.

[8.25] Griffith, R. L., Wright, J. T., Maldonado, J., Povich, M. S., Sigurđsson, S. and Mullan, B. (2015) The Ĝ Infrared Search for Extraterrestrial Civilizations with Large Energy Supplies. III. The Reddest Extended Sources in WISE. *The Astrophysical Journal Supplement Series* 217, #25.

[8.26] Grimaldi, C. (2017) Signal coverage approach to the detection probability of hypothetical extraterrestrial emitters in the Milky Way. *Scientific Reports* 7, #46273.

[8.27] Grimaldi, C., Marcy, G. W., Tellis, N. K. and Drake, F. (2018) Area Coverage of Expanding E.T. Signals in the Galaxy: SETI and Drake's N. *Publications of the Astronomical Society of the Pacific* 130, #054101.

[8.28] Grimaldi, C. and Marcy, G. W. (2018) Bayesian approach to SETI. *Proceedings of the National Academy of Sciences* 115, E9755-E9764.

[8.29] Hart, M. H. (1975) Explanation for the Absence of Extraterrestrials on Earth. *Quarterly Journal of the Royal Astronomical Society* 16, 128-135.

[8.30] Horvath, J. E. and Galante, D. (2012) Effects of high-energy astrophysical events on habitable planets. *International Journal of Astrobiology* 11, 279-286.

[8.31] Howard, A. W., Horowitz, P., Wilkinson, D. T., Coldwell, C, M., Groth, E. J., Jarosik, N., Latham, D. W., Stefanik, R. P., Willman, A. J., Jr., Wolff, J. and Zajac, J. M. (2004) Search for Nanosecond Optical Pulses from Nearby Solar-Type Stars. *The Astrophysical Journal* 613, 1270-1284.

[8.32] Isaacson, H., Siemion, A. P. V., Marcy, G. W., Lebofsky, M., Price, D. C., MacMahon, D., Croft, S., DeBoer, D., Hickish, J., Werthimer, D., Sheikh, S., Hellbourg, G. and Enriquez, J. E. (2017) The Breakthrough Listen Search for Intelligent Life: Target Selection of Nearby Stars and Galaxies. *Publications of the Astronomical Society of the Pacific* 129, #054501

[8.33] Kardashev, N. S. (1964) Transmission of Information by Extraterrestrial Civilizations. *Soviet Astronomy* 8, 217-221.

[8.34] Kiang, N. Y., Domagal-Goldman, S., Parenteau, M. N., Catling, D. C., Fujii, Y., Meadows, V. S., Schwieterman, E. W. and Walker, S. I. (2018) Exoplanet Biosignatures: At the Dawn of a New Era of Planetary Observations. *Astrobiology* 18, 619-629.

[8.35] Kim, S. H. and Whitt, W. (2013) Estimating waiting times with the time-varying Little's law. *Probability in the Engineering and Informational Sciences* 27, 471-506.

[8.36] Kuhn, J. R., and Berdyugina, S. V. (2015) Global warming as a detectable thermodynamic marker of Earth-like extrasolar civilizations: the case for a telescope like Colossus. *International Journal of Astrobiology* 14, 401-410.

[8.37] Lazio, T. J. W., Tarter, J. C. and Wilner, D. J. (2004) The cradle of life. *New Astronomy Reviews* 48, 985-991.

[8.38] Lin, H. W., Abad, G. G. and Loeb, A. (2014) Detecting Industrial Pollution in the Atmospheres of Earth. *The Astrophysical Journal Letters* 792, #L7.

[8.39] Lineweaver, C. H., Fenner, Y., and Gibson, B. K. (2004) The galactic habitable zone and the age distribution of complex life in the Milky Way. *Science* 303, 59-62,

[8.40] Lingam, M. and Loeb, A. (2017) Fast Radio Bursts from Extragalactic Light Sails. *The Astrophysical Journal Letters* 837, #L23.

[8.41] Little, J. D. C. (1961) A proof for the queuing formula. *Operations Research* 9, 383-387.

[8.42] Little, J. D. C., and Graves, S. C. (2008) Little's law. In *Building Intuition: Insights from Basic Operations Management: Models and Principles*. D. Chhajed and TJ. Lowe (eds.) 115, 81-100.

[8.43] Loeb, A. and Turner, E. L. (2012) Detection Technique for Artificially Illuminated Objects in the Outer Solar System and Beyond. *Astrobiology* 12, 290-294.

[8.44] Loeb, A., Batista, R. A., and Sloan, D. (2016) Relative likelihood for life as a function of cosmic time. *Journal of Cosmology and Astroparticle Physics* 8, 40.

[8.45] Lubin, P. (2016) The search for directed intelligence, *REACH*, 1, 20-45.

[8.46] NASA and the Search for Technosignatures: A Report from the NASA Technosignatures Workshop, eprint arXiv:1812.08681.

[8.47] Petigura, E. A., Howard, A. W., and Marcy, G.W. (2013) Prevalence of Earth-size planets orbiting Sun-like stars. *Proceedings of the National Academy of Sciences* 110, 19273–19278.

[8.48] Price, D. C., Enriquez, J. E., Brzycki, B. , Croft, S., Czech, D., DeBoer, D., DeMarines, J., Foster, G., Gajjar, V., Gizani, N., Hellbourg, G., Isaacson, H., Lacki, B., Lebofsky, M., MacMahon, D. H. E., Pater, I., Siemion, A. P. V., Werthimer, D., Green, J. A., Kaczmarek, J. F., Maddalena, R. J.. Mader, S. , Drew, J. and Worden S. P. (2020) The breakthrough listen search for intelligent life: observations of 1327 nearby stars over 1.10–3.45 GHz. *The Astronomical Journal* 159, #86.

[8.49] Schneider, J., Léger, A., Fridlund, M., White, G. J., Eiroa, C., Henning, T., Herbst, T., Lammer, H., Liseau, R., Paresce, F., Penny, A., Quirrenbach, A., Röttgering, H., Selsis, F., Beichman, C., Danchi, W., Kaltenegger, L., Lunine, J., Stam, D. and Tinetti G. (2010) The Far Future of Exoplanet Direct Characterization. *Astrobiology* 10, 121-126.

[8.50] Siemion, A., Benford, J., Cheng-Jin, J., Chennamangalam, J., Cordes, J. M., Falcke, H. D. E., Garrington, S. T., Garrett, M. A., Gurvits, L., Hoare, M., Korpela, E., Lazio, J., Messerschmitt, D., Morrison, I., O'Brien, T., Paragi, Z., Penny, A., Spitler, L., Tarter, J. and Werthimer, D. (2015) Searching for extraterrestrial intelligence with the Square Kilometre Array. In *Advancing Astrophysics with the Square Kilometre Array (AASKA14)*, eds Bourke, T. L. et al. (International School for Advanced Studies, Trieste, Italy), Vol 215, no. 116.

[8.51] Socas-Navarro, E. (2018) Possible Photometric Signatures of Moderately Advanced Civilizations: The Clarke Exobelt. *The Astrophysical Journal* 855, #110.

[8.52] Spiegel, D. S. and Turner, E. L. (2012) Bayesian analysis of the astrobiological implications of life's early emergence on Earth. *Proceedings of the National Academy of Sciences* 109, 395-400.

[8.53] Tarter, J. (2001) The Search for Extraterrestrial Intelligence (SETI). *Annual Review of Astronomy and Astrophysics* 35, 511-548.

[8.54] Tellis, N. K. and Marcy, G. W. (2017) A Search for Laser Emission with Megawatt Thresholds from 5600 FGKM Stars. *The Astronomical Journal* 153, #251.

[8.55] Ward, P. D. and Brownlee, D. (2000) *Rare Earth: Why Complex Life Is Uncommon in the Universe*, Copernicus Springer-Verlag, New York.

[8.56] Webb, S. (2002) *If the Universe Is Teeming with Aliens ... where is everybody?: Fifty Solutions to the Fermi Paradox and the Problem of Extraterrestrial Life*, Springer, New York, USA.

[8.57] Wright, J. T., Cartier, K. M. S., Zhao, M., Jontof-Hutter, D. and Ford, E. B. (2015) The Search for Extraterrestrial Civilizations with Large Energy Supplies. IV. The Signatures and Information Content of Transiting Megastructures. *The Astrophysical Journal* 816, #17.
[8.58] Wright, J. T., Kanodia, S. and Lubar, E. (2018) How Much SETI Has Been Done? Finding Needles in the n-dimensional Cosmic Haystack. *The Astronomical Journal* 156, #260.

9

Economics and Technosignatures: New Connections

Anamaria Berea

George Mason University, Fairfax, VA, USA

Abstract
In this chapter I explore the many ways in which economics and technosignatures can bring forth fruitful connections between the two fields, from methods to findings and analyses. I also comment on the larger field of astrobiology. I note how social sciences and particularly economics can shed some light or point directions towards finding answers to particular questions regarding social behavior, cooperation and competition, complex systems, and so on in living systems on Earth that we are also likely to find elsewhere.

Keywords: Economics, technosignatures, astrobiology, complex systems, competition and cooperation, social behavior

9.1 The Different Faces of Economics

The extensions of astrobiology into social sciences and the implications of social sciences for astrobiology are two research areas where both disciplines can draw very useful benefits from each other [9.56, 9.76, 9.45]. For example, social science can inform astrobiology regarding the evolution of social behavior in biological systems, from microbes to evolved organisms [9.8, 9.10, 9.43, 9.54, 9.58, 9.82], or the effects of social behavior on the planetary environment at large [9.66, 9.36, 9.71]. At the same

Email: aberea@gmu.edu

Anamaria Berea (ed.) Technosignatures for Detecting Intelligent Life in Our Universe: A Research Companion, (149–160) © 2022 Scrivener Publishing LLC

time, social science can back up astrobiology with respect to the effects and implications of astrobiological findings in the evolution of our society and social behavior response [9.77, 9.56]. Not least, methodologies from social science, such as computational social science and complex social systems, can help explore astrobiology hypotheses in novel ways. Economics is obviously a social science and, for example, current economic research on sustainable systems or ecological economics research on the coevolution of social, technological systems and the environment can shed new light on macro patterns of planetary evolution. While biological and social systems on our planet are closely linked and often studied together, an extension of these studies into the astrobiology field seems intuitive. But there is still a very large gap between economics as a social science and astrobiology, and in this paper I would like to show how much potential for collaboration and scientific and scholarly advancement can exist between both these disciplines once they start "talking" to each other.

Some fields in economics that have addressed some of the questions that astrobiology is also trying to answer are ecological economics, complex systems economics, energy economics, computational economics, behavioral economics, and space/GIS economics, to name only a few. In this proposal, I briefly outline how we can align these two disciplines, thus opening new opportunities for cataloguing and referencing universal phenomena for biosignatures and technosignatures research, as well as creating new opportunities for interdisciplinary research as an entirely new research program.

The following table is a guideline/roadmap that includes only a few examples, for illustration purposes, of how some of the current research in economics can inform astrobiology. We can collect such examples of current economic research and findings into a comprehensive account of insights with importance for astrobiology.

Research reference/areas	Biological systems	Social systems	Proto economic systems	Technological systems	Branch of economics
Sustainability/Failures and Crises	Sustainable habitats, …	Eusocial insects, Swarming behavior, …	Trade routes vs. wars, Emergence of cities, …	Financial Crises, Political Cycles, …	Macroeconomics, History of Economics, Complex Systems Economics, Ecological Economics
Energy Expenditures (Costs and Benefits)	Metabolism, …	Animal constructions (bridges, nests, mounds), …	Ships at sea, Mesopotamian granaries, …	Hydro plants, Nuclear plants, Solar plants, …	Macroeconomics, Energy Economics, Ecological Economics
Coevolution with Environment	Mutation and selection, …	Cetacean cultures differentiation Atlantic vs. Pacific, …	Ingroup/Outgroup dynamics, …	Human habitats, …	Complex Systems Economics, Computational Economics
Informational Expenditures (Costs and Benefits)	Gene signaling, …	Animal communication, Animal migration patterns, …	Trade routes, invention of writing and accounting, …	Digital communication, Human migration patterns, …	Microeconomics, Public Choice Economics, Space/GIS economics
Universal Laws/Contextual Phenomena	Entropy and efficiency, …	Free riding behavior, …	Power laws of conflict magnitude, …	Zipf law in language vs. small world networks, …	Computational Economics, Complex Systems Economics, Philosophy of Economics

9.2 Economics in the Context of NASA Astrobiology Roadmap and the Drake Equation

The NASA Astrobiology Roadmap identifies the importance of interdisciplinarity and multiple approaches to a subject of such immense importance as understanding life in the universe. Perhaps a humanistic discipline such as economics does not come to mind immediately as one of the main fields in terms of astrobiology ramifications. But so much research from the past few years in the field of astrobiology has been pointing towards the need for a deeper understanding of phenomena such as: the coevolution of environment and life in general [9.20, 9.36], the universality of energy consumption in biological and technological systems [9.4, 9.26, 9.33, 9.50], the universality of information exchange in biological and social systems [9.24, 9.30] or the sustainability of ecosystems [9.27, 9.28, 9.31], to name just a few. All these are extremely important for understanding the potential for finding biosignatures and even more so, for the technosignatures research [9.19, 9.56]. Therefore, questions about sustainability, energy consumption, biological-social-technological coevolution, and universal laws from graph and statistical regularities are just a few examples where economics and the social sciences have been providing some answers in the past few decades [9.2, 9.5, 9.6, 9.4, 9.14, 9.18, 9.23, 9.32, 9.37, 9.44, 9.48, 9.35, 9.25, 9.59, 9.68, 9.65].

The Drake equation is another roadmap, a measurable one, that has been guiding us on estimating the possibility of life and intelligent life on other planets [9.57]. But particularly, two factors in the Drake equation can be better estimated using economics as we know it today, f_c and L:

> f_c = the fraction of civilizations that develop a technology that releases detectable signs of their existence into space
> L = the length of time for which such civilizations release detectable signals into space

In this respect, economics can point towards those universal laws that can be found both in natural and social systems, thus making these laws more likely to be found elsewhere in the universe. Examples of such laws are random walk behavior (statistical regularities) [9.14] or power law distributions of various collective phenomena (scaling laws) [9.63, 9.59, 9.69], network or graph specific laws (i.e., small world networks or fractal laws) [9.44, 9.42, 9.55, 9.80], and throughout this research one of the main tasks would be to catalogue all these universal economic laws relevant for the astrobiology field in one place of reference.

The rise and fall of civilizations has been widely studied both in the humanities [9.67, 9.53, 9.51, 9.39, 9.41, 9.62, 9.66] and the sciences [9.21, 9.29], and particularly by historians in the humanities [9.16] and by the computational social scientists in the sciences [9.7, 9.11]. However, when we look at civilizations on Earth, past or present, we immediately recognize the strong link between the dynamics of a civilization and its economic evolution [9.34].

In other words, by exploring the current status quo in the field of civilization sustainability that economists, historians and history of economics researchers have been exploring, we can get a better understanding and estimation of f_c, while by exploring the current status quo in the field of the history of technology and the technological sustainability within a civilization, we can get to a better estimation of L.

The other side of the coin (or the null hypothesis) of sustainability in biological, social, economic and technological systems is systems failure [9.70]. Failures and crises have also been studied by economists, particularly macroeconomists, in terms of not only country-level economic crises, but also with deeper implications, such as evolutionary failures, collective and social behavior failures, technological failures and communication failures that have led to crises or the demise of countries [9.40, 9.38, 9.49, 9.51]. An outlook on the findings of the macroeconomists will help us better understand sustainability and its limits in the context of our civilizations, past and present.

9.3 Economics in the Context of Biosignatures vs. Technosignatures Research

One difference between living and non-living systems is collective action—individuals that act together towards an expected goal or individuals that act alone but the emergent result is that of a collective, with a result larger than the sum of its parts (i.e., swarms). In non-living systems, such a property of individual actions morphing into another phenomenon does not exist [9.63]. In living organisms, this property of the emergent collective action of a system has been widely researched by economists and social biologists, who have discovered examples of social behavior in many species, such as markets of exchange in bacteria, competition in slime molds, free riding in bees and primates, and many more [9.81, 9.78, 9.75, 9.71, 9.73]. Therefore, an account of the roots of collective action in economics and complex systems will undoubtedly help us to understand the framework of biosignatures at large, as a signal of a collective or emergent phenomenon that is different from the sum of its parts.

The problem of collective action is more fundamental than tracing the roots of social behavior in biology. The larger problem that scientists are trying to solve is with respect to scaling from one to two and from two to many [9.1]. This stems from a larger, epistemological problem, where scientists from physics, information theory, network science, chemistry, biology and the social sciences have been trying to answer the fundamental question of bonding/dissociation of entities in the universe, from cells to societies [9.55, 9.61, 9.47, 9.43, 9.24, 9.17]. Cataloguing and referencing the current findings in the philosophy of economics in this respect will help deepen our understanding of what to expect in terms of biosignature/biosignatures in the universe.

The problem of collective action is also highly related to problems of competition and cooperation, or adaptation and selection, that we can find in all living organisms [9.47, 9.7]. Competition and cooperation have been widely studied by economists and the view of how humans cooperate or compete both in society and with the environment, comparatively with how other species do the same within their groups and environments, will help us better understand the dynamics of selection and adaptation and its effect on what constitutes a biosignature of a group versus another [9.52, 9.64, 9.79].

All organisms use costs and benefits, and the areas where we see this the most in all living organisms are in signaling/communication, metabolism and in mating or creating value for the future generations [9.58, 9.60, 9.46, 9.15, 9.12, 9.13]. Unlike economists, organisms do not use mathematics to understand cost and benefit, but the principles are the same: any individual organism will try to get more benefit and spend less in terms of energy or information with each of its actions. While cost and benefit analyses are used by scientists that study metabolisms, gene signaling or animal communication, the theoretical foundations for the cost and benefit research both in energy and information have been laid down by economists and they can inform further the universality of such an approach beyond life on Earth.

Macroeconomic energy studies have consistently shown how our civilization has increased its energy consumption consistently with its transition from agrarian to industrial and then to informational age [9.3, 9.74]. There is a clear connection between the technological advancement of a society and its energy consumption, and this can be seen in the patterns of lights and cities at night or in the analysis of satellite images to understand the impact of certain economic policies over others, as studied by the field of Space/GIS economics [9.9].

The reverse aspect of the coevolution of living systems and the environment is the study of the effects the environment has had on living systems, such as migration patterns, both in humans and other species [9.72]. While some human migrations are not only due to the environment, but also to economic/society factors, migrations are a planetary macro phenomenon that can be observed from space as well. Additionally, habitats (human or otherwise) with less environmental change are more or less likely to be sustainable as well and ecological economics can provide us with the answers for when this happens and when it does not.

9.4 Economic Methodology and Epistemology into Astrobiology

Economics has long sought to find general principles and laws of human behavior in the same vein as the hard sciences, such as physics and biology have. While doing so, many economists have discovered principles of social behavior that expand beyond the human societies, into other species. They have also discovered what is contextual as well as what is universal about human behavior and placed the understanding of humanity in the context of space and time, or removed humanity from the context of space and time. By doing so, methodologically, economics has achieved deeper insights into comparative studies (between different contexts).

Economics helps not only with understanding human behavior, but with understanding how principles of human behavior are fundamental in biology, such as signaling, free riding, individual vs. collective, and game theory. These are broad applications that have found grounds in theoretical biology as well.

Computational economics is great at simulating the rise and fall of civilizations, energy dynamics, and segregation patterns and, in general, is a field of economics specialized in using computer simulations to study the emergence of macro patterns from micro behaviors. Undoubtedly, many of the results in this field will help inform astrobiology with how to design experiments for the study of biosignatures and technosignatures. In fact, some of these studies have already preliminarily been done at the NASA Frontier Development Lab in 2018, where data scientists and astrobiologists worked together to explore what is universally possible for life.

Ecological economics can provide answers to the effects of technology and social activity into the environment and vice versa, as well as with sustainability in the context of the environment. Macroeconomics can provide answers with respect to the sustainability in terms of economic and technological evolution, as well as answers to the likely causes of crises and failures. Space and GIS economics, a more recent branch of economics, can show us how satellite imagery analysis is an indicator for energy or information expenditures and the different effects of sustainability and environment conditions on different parts of the planet.

And last, but not least, economics can help us understand the development and future of the astrobiology field today, given the current economic constraints and the impact of the field in society.

References

[9.1] The National Academies Keck Futures Initiative (ed.), Collective Behavior – From Cells to Societies. The National Academies Keck Futures Initiative (2015).

[9.2] Alperovitz, G. (2009). The rise of the new economy movement, http://www.alternet.org/story/155452/the_rise_of_the_new_economy_movement

[9.3] Allenby, B. and Richards, D. (1994). *The greening of industrial ecosystems.* The Advisory Committee on Industrial Ecology and Environmentally Preferable Technology, National Academy of Engineering (Apr 1, 1994). Washington, DC: National Academy Press.

[9.4] Arthur, W.B. (1990). Positive feedback in the economy. *Scientific American,* 262(2), 92-99.

[9.5] Arrow, K.J. and Debreu, G., (1954). Existence of an equilibrium for a competitive economy. *Econometrica: Journal of the Econometric Society,* pp.265-290.

[9.6] Akerlof, G.A. (1970). The market for lemons: quality uncertainty and the market mechanism. *Q. J. Econ.,* 84(3), 488–500.

[9.7] Axelrod, R. (1997). The Complexity of Cooperation: Agent-Based Models of Competition and Collaboration (Princeton University Press, Princeton).

[9.8] Axelrod, R., & Hamilton, W. D. (1981). The evolution of cooperation. *Science, 211*(4489), 1390-1396.

[9.9] Bateman, I. J., Lovett, A. A., Brainard, J. S., & Pearce, D. W. (2003). *Applied environmental economics: A GIS approach to cost-benefit analysis.* Cambridge University Press.

[9.10] Bateson, M., Desire, S., Gartside, S. E., & Wright, G. A. (2011). Agitated honeybees exhibit pessimistic cognitive biases. *Current Biology, 21*(12), 1070-1073.

[9.11] Berea, A. (2010). Trade as a premise for social complexity.
[9.12] Berea, A. (2018). How does intelligent life communicate? An atlas of communication evolution based on a unified database of evidence. DAI workshop.
[9.13] Birchler, U. W., & Bütler, M. (2007). Information economics Routledge advanced texts in economics and finance (pp. xxiii, 462 p.).
[9.14] Blaug, M. and Lloyd, P. (2010). *Famous figures and diagrams in economics*. New York: Edward Elgar.
[9.15] Bradbury, J. W., & Vehrencamp, S. L. (2000). Economic models of animal communication. *Animal behaviour, 59*(2), 259-268.
[9.16] Braudel, F. (1995). *A history of civilizations* (p. 178). New York: Penguin Books.
[9.17] Braman, S. (2006). The micro- and macroeconomics of information. *Annual review of information science and technology, 40*(1), 3-52.
[9.18] Brue, S., & Randy G. (2012). *The evolution of economic thought*. Cengage Learning (online), 2012.
[9.19] Cabrol, N. A. (2016). Alien mindscapes—a perspective on the Search for Extraterrestrial Intelligence.
[9.20] Cabrol, N. A. (2018). The Coevolution of Life and Environment on Mars: An Ecosystem Perspective on the Robotic Exploration of Biosignatures.
[9.21] Currie, T. E., Greenhill, S. J., Gray, R. D., Hasegawa, T., & Mace, R. (2010). Rise and fall of political complexity in island South-East Asia and the Pacific. *Nature, 467*(7317), 801.
[9.22] Des Marais, D. J., Nuth III, J. A., Allamandola, L. J., Boss, A. P., Farmer, J. D., Hoehler, T. M., ... & Spormann, A. M. (2008). The NASA astrobiology roadmap. *Astrobiology, 8*(4), 715-730.
[9.23] Ekelund, R. B., and Herbert, F. 2013. A history of economic theory and method (6th edition). Long Grove, Il., USA: Waveland Press, Inc.
[9.24] Emmeche, C. (1994). The computational notion of life. *Theoria: An International Journal for Theory, History and Foundations of Science*, 1-30.
[9.25] Erdos, P., & Rényi, A. (1960). On the evolution of random graphs. *Publ. Math. Inst. Hung. Acad. Sci, 5*(1), 17-60.
[9.26] Fath, B. (2007). Network mutualism: Positive community level relations and ecosystems: *Ecological Modelling*, 208, 56-67.
[9.27] Fath B.D., Dean C.A., Katzmair H. (2015). Navigating the adaptive cycle: An approach to managing the resilience of social systems. *Ecology and Society*. 20(2): 24.
[9.28] Fischer-Kowalski, M. and Hüttler, W. (1998). Society's metabolism: the intellectual history of material flow analysis – Part I: 1860-1970. *Journal of Industrial Ecology*, **2(1)**, 61-78. Part II: 1970-1998. **2(1)**, 107–137.
[9.29] Fernández-Armesto, F. (2001). *Civilizations: culture, ambition, and the transformation of nature*. Simon and Schuster.
[9.30] Friedkin, N. E. (1982). Information flow through strong and weak ties in intraorganizational social networks. *Social networks, 3*(4), 273-285.

[9.31] Goerner, S., Dyck, R. and Lagerroos, D. (2008). *The new science of sustainability: Building a foundation for great change.* Chapel Hill, NC: Triangle Center for Complex Systems, distributed by Gabriola Island, BC, Canada: New Society Publishers.

[9.32] Goerner, S., Lietaer, B., and Ulanowicz, R. (2009). Quantifying economic sustainability: Implications for free enterprise, theory, policy and practice. *Ecological Economics*, 69(1), 76-81.

[9.33] Goerner, S. J., Fath, B., and Fiscus, D. submitted. Using energy network science (ENS) to connect resilience with the larger story of systemic health and development. *Ecological Economics*. Submitted.

[9.34] Gunderson, L. and Holling, C.S. (2001) *Panarchy: Understanding Transformations in Human and Natural Systems.* New York: Island Press.

[9.35] Granovetter, M. (1985). Economic action and social structure: The problem of embeddedness. *American journal of sociology, 91*(3), 481-510.

[9.36] Grinspoon, D. (2009). *Lonely planets: The natural philosophy of alien life.* Harper Collins.

[9.37] Hamilton, D. (1962). Why is Institutional economics not institutional? *The American Journal of Economics and Sociology.* 21(3), July, 309-317.

[9.38] Holland, J. H. (2012). *Signals and boundaries: Building blocks for complex adaptive systems.* Mit Press.

[9.39] Jacobs, J. (1961/1989). *The death and life of great American cities.* New York: Vintage books.

[9.40] Jacobs, J. (1969). *The economy of cities.* New York: Vintage Books, a Division of Random House.

[9.41] Jacobs, J. (1984). *Cities and the Wealth of Nations.* New York: Random House.

[9.42] Jackson, M. O. (2008). Social and Economic Networks, Princeton, NJ: Princeton Univ.

[9.43] Kauffman, S. A. (1993). *The origins of order: Self-organization and selection in evolution.* Oxford University Press, USA.

[9.44] Mandelbrot, B. B. (2006). *The misbehavior of markets: A fractal view of financial turbulence.* New York: Basic Books.

[9.45] McKay, C. P. (2013). Astrobiology and Society. *Encountering Life in the Universe: Ethical Foundations and Social Implications of Astrobiology*, 158.

[9.46] Miller, M. B., & Bassler, B. L. (2001). Quorum sensing in bacteria. *Annual Reviews in Microbiology, 55*(1), 165-199.

[9.47] Miller, J. H., & Moser, S. (2004). Communication and coordination. *Complexity, 9*(5), 31-40.

[9.48] North, D. (1991). Institutions. *Journal of Economic Perspectives.* 5(1): 640-655.

[9.49] Odum, H.T. (1988). Self-Organization, Transformity, and Information. *Science*, 242, 1132.

[9.50] Odum, H. T. (2007). *Environment, power and society for the 21st century: The hierarchy of energy*, New York: Columbia University Press.

[9.51] Ostrom, E. (1990). *Governing the commons: The evolution of institutions for collective action*. London: Cambridge University Press.

[9.52] Ostrom, E. (2010). Beyond markets and states: polycentric governance of complex economic systems. *American economic review, 100*(3), 641-72.

[9.53] Ostrom, V. (1997). *The meaning of democracy and the vulnerability of democracies: A response to Tocqueville's challenge*. University of Michigan Press.

[9.54] Pattee, H. H. (2015). Cell phenomenology: The first phenomenon. *Progress in biophysics and molecular biology, 119*(3), 461-468.

[9.55] Prigogine, I. (1972). Thermodynamics of evolution. *Physics Today*, December, 1972.

[9.56] Race, M., Denning, K., Bertka, C. M., Dick, S. J., Harrison, A. A., Impey, C., & Mancinelli, and Workshop Participants, R. (2012). Astrobiology and society: building an interdisciplinary research community. *Astrobiology, 12*(10), 958-965.

[9.57] Sagan, C., & Drake, F. (1975). The search for extraterrestrial intelligence. *Scientific American, 232*(5), 80-89.

[9.58] Searcy, W. A., & Nowicki, S. (2005). *The evolution of animal communication: reliability and deception in signaling systems*. Princeton University Press.

[9.59] Schelling, T. C. (2006). *Micromotives and macrobehavior*. WW Norton & Company.

[9.60] Scott-Phillips, T. C. (2008). On the correct application of animal signalling theory to human communication. In *The evolution of language* (pp. 275-282).

[9.61] Simon, H. A. (2002). Near decomposability and the speed of evolution. *Industrial and corporate change, 11*(3), 587-599.

[9.62] Sen, Amartya. 1999. *Development as freedom*. New York: Oxford University Press.

[9.63] Simon, H. A. (1996). *The sciences of the artificial*. MIT press.

[9.64] Skyrms, B. (2010). *Signals: Evolution, learning, and information*. Oxford University Press.

[9.65] Stigler, G. J. (1961). The economics of information. *Journal of political economy, 69*(3), 213-225.

[9.66] Stiglitz, J. and Greenwald, B. (2014). Creating a learning society: A new approach to growth, development, and social progress. New York: Columbia University

[9.67] Smith, A. (1776/1991). *The wealth of nations*. Amherst, New York: Prometheus Books.

[9.68] Smith, V. L. (1976). Experimental economics: Induced value theory. *The American Economic Review*, 66(2): 274-279.

[9.69] Taleb, N. (2010). *The black swan: The impact of the highly improbable*. New York: Random House.

[9.70] Thom, R. (2018). *Structural stability and morphogenesis*. CRC Press.

[9.71] Ulanowicz, R. (2009). *A third window: Natural life beyond Newton and Darwin.* West Conshohocken, PA (USA): Templeton Foundation Press.

[9.72] Ulanowicz, Robert E. (2012). *Growth and development: ecosystems phenomenology*. Springer Science & Business Media, 2012.

[9.73] Ulanowicz, R. E., Goerner, S. J., Lietaer, B., & Gomez, R. (2009). Quantifying sustainability: resilience, efficiency and the return of information theory. *Ecological complexity*, *6*(1), 27-36.

[9.74] Ulanowicz, R. E., & Hannon, B. M. (1987). Life and the production of entropy. *Proc. R. Soc. Lond. B*, *232*(1267), 181-192.

[9.75] Van Laerhoven, F. and Ostrom, E. (2007). Traditions and trends in the study of the Commons *International Journal of the Commons, 1*(1), October 2007, pp. 3-28.

[9.76] Vakoch, D. A. (2013). *Astrobiology, History, and Society*. Springer.

[9.77] Vakoch, D. A., & Albert, A. (Eds.). (2013). *Civilizations beyond Earth: extraterrestrial life and society*. Berghahn Books.

[9.78] Waddington, C. H. (1957). The strategy of the genes. A discussion of some aspects of theoretical biology. With an appendix by H. Kacser. *The strategy of the genes. A discussion of some aspects of theoretical biology. With an appendix by H. Kacser.*

[9.79] Walker, S. I., & Davies, P. C. (2013). The algorithmic origins of life. *Journal of the Royal Society Interface*, *10*(79), 20120869.

[9.80] Watts, D. J., & Strogatz, S. H. (1998). Collective dynamics of 'small-world'networks. *nature*, *393*(6684), 440.

[9.81] Werner, G. D., Strassmann, J. E., Ivens, A. B., Engelmoer, D. J., Verbruggen, E., Queller, D. C., ... & Kiers, E. T. (2014). Evolution of microbial markets. *Proceedings of the National Academy of Sciences*, *111*(4), 1237-1244.

[9.92] Whitehead, H. & Rendell, L. (2014). The Cultural Lives of Whales and Dolphins. University of Chicago Press, Chicago.

Index

Also of Interest

Check out these published and forthcoming related titles from Scrivener Publishing

Origin of Life via Archaea
By Richard Gordon and Natalie K. Gordon
Forthcoming 2023.

Pathways to the Origin and Evolution of Meanings in the Universe
Edited by Alexei Sharov, Richard Gordon and George Mikhailovsky
Forthcoming 2023. ISBN 978-1-119-86509-4

Conflicting Models for the Origin of Life
Edited by Stoyan K. Smoukov, Joseph Seckbach, Richard Gordon
Forthcoming 2022. ISBN 978-1-119-55552-0

Technosignatures for Detecting Intelligent Life in Our Universe: A Research Companion
Edited by Anamaria Berea
Published 2022. ISBN 978-1-119-64040-0

Terraforming Mars
Edited by Martin Beech, Joseph Seckbach and Richard Gordon
Published 2022. ISBN 978-1-119-76196-9

Planet Formation and Panspermia
New Prospects for the Movement of Life through Space
Edited by B. Vukotic, Joseph Seckbach and Richard Gordon
Published 2021. ISBN 978-1-119-64039-4

Astrobiology
Science, Ethics, and Public Policy
Edited by Octavio A. Chon Torres, Ted Peters, Joseph Seckbach and Richard Gordon
Published 2021. ISBN 978-1-119-71116-2

Extremophiles as Astrobiological Models
Edited by Joseph Seckbach and Helga Stan-Lotte
Published 2020. ISBN 978-1-119-59168-9

www.scrivenerpublishing.com

Printed and bound by CPI Group (UK) Ltd, Croydon, CR0 4YY
17/09/2026
14964246-0001